T0331599

Simplicial Complexes in Complex Systems

In Search for Alternatives

Simplicial Complexes in Complex Systems
In Search for Alternatives

YI ZHAO

Harbin Institute of Technology Shenzhen, China

SLOBODAN MALETIĆ

Institute of Nuclear Sciences Vinca,
University of Belgrade, Serbia

NEW JERSEY · LONDON · SINGAPORE · BEIJING · SHANGHAI · HONG KONG · TAIPEI · CHENNAI · TOKYO

Published by

World Scientific Publishing Co. Pte. Ltd.

5 Toh Tuck Link, Singapore 596224

USA office: 27 Warren Street, Suite 401-402, Hackensack, NJ 07601

UK office: 57 Shelton Street, Covent Garden, London WC2H 9HE

British Library Cataloguing-in-Publication Data

A catalogue record for this book is available from the British Library.

Originally published in English by Harbin Institute of Technology Press Ltd.

Copyright © Harbin Institute of Technology Press Ltd., 2018

This edition is distributed worldwide by World Scientific Publishing Co. Pte. Ltd., except China.

SIMPLICIAL COMPLEXES IN COMPLEX SYSTEMS
In Search for Alternatives

Copyright © 2021 by World Scientific Publishing Co. Pte. Ltd.

All rights reserved. This book, or parts thereof, may not be reproduced in any form or by any means, electronic or mechanical, including photocopying, recording or any information storage and retrieval system now known or to be invented, without written permission from the publisher.

For photocopying of material in this volume, please pay a copying fee through the Copyright Clearance Center, Inc., 222 Rosewood Drive, Danvers, MA 01923, USA. In this case permission to photocopy is not required from the publisher.

ISBN 978-981-122-631-1 (hardcover)
ISBN 978-981-122-632-8 (ebook for institutions)
ISBN 978-981-122-633-5 (ebook for individuals)

For any available supplementary material, please visit
https://www.worldscientific.com/worldscibooks/10.1142/11991#t=suppl

Typeset by Stallion Press
Email: enquiries@stallionpress.com

Printed in Singapore

Preface

A tremendous impact of complex systems on our lives recruited an army of researchers devoted to better understanding properties of complex systems, which would eventually lead to controlling and predicting their behavior (of complex systems, not researchers). Fast development of the research in complex systems has led to an explosion of publications and results which shed new light on the multitude of real world phenomena, and it is followed with the development of mathematical methods for tackling the emerging problems. Nevertheless, the development of mathematical methods does not have to be always used for solving problems, but can serve as a tool for extracting not so obvious properties, like building the Mars rover to discover hidden features which are not detected with previous tools.

One goal of writing a book is to introduce an alternative and rich mathematical framework, grounded on building simplicial complexes, for tackling the problems emerging from the complex systems research. Yet another goal is to introduce alternative mathematical tools, based on the relationships between simplices (to say an alternative Mars rover), for discovering new, rather disguised, features of complex systems. The motivation which led us to write this book is actually twofold. First, there is the lack of literature which covers very broad applications of simplicial complexes, hence not restricted only to the topological data analysis or qualitative structural analysis. And second, we want to introduce to the reader algebraic topology concepts in rather intuitive way, which (at least we hope) is sufficient for simple applications, and good start base for further upgrading the knowledge.

In order to clarify abstract concepts to the reader, the book abounds with examples illustrated in figures, and as the reading progresses the examples become more complicated. After introductory chapter, where we put the content of the book in broader research context, we proceed with chapters dedicated to the definitions of simplicial complexes, their properties, and building them from data. Finally we conclude with the illustration of the potential real applications of simplicial complexes, and we emphasize the possible applications in the aeronautical science, as an example of how the versatility of the algebraic topology concepts can contribute to the specific research field.

The authors are grateful to the Harbin Institute of Technology, Shenzhen, for providing the favorable working conditions during the course of writing the book. Also the authors acknowledge supports from the Innovative Project of Shenzhen, China under Project No. KQJSCX20180328165509766 and the Natural Science Foundation of Guangdong Province, China under Grant No. 2020A1515010812 and 2021A1515011594.

Fruitful discussions with Dr. Nikola Mirkov from the Institute of Nuclear Sciences Vinča helped in clarifying the applicability opportunities in the aeronautical science of mathematical concepts and tools which are introduced in the book. Y. Z. is grateful to Prof. Michael Small for introducing the world of complex systems to him. S. M. expresses an appreciation to Prof. Milan Rajković for collaboration, and for initiation of the research in simplicial complexes, as well as encouraging the pursuit in versatile applications of simplicial complexes.

The authors also would like to thank the postgraduate students (Mr. Andu Zhou, Ms. Min Ma, Mr. Xinyu Han, and Mr. Dong Wang) in our team for their proof and reading comments on this book. Their works made it eminently readable and especially suitable to fresh readers.

March 29, 2021

Contents

1

Introduction

When we step into an unknown field or forest, and walk through it, we are becoming familiar with the terrain very roughly at the beginning, and then with more details. This book serves as an initial guidance over the field of complex systems seen through the lens of simplicial complexes. Although mathematical origin of simplicial complexes carries with itself a high level of abstraction, we will try to make it accessible to the reader having no prior knowledge about either algebraic topology or simplicial complexes. In this introductory chapter we will give a short walk through the field of complex systems, and we will introduce basic notions which are enough to get a rough picture about the landscape that properties and concepts shape. We will end up by an overview of the successful applications of simplicial complexes as mathematical tools. At the very beginning we want to stress that we believe that picture can be more helpful in introducing and understanding new concepts rather than long explanations, and whence throughout the text we will try wherever it is appropriate to illustrate and support our assertions with figures, at the same time trying not to lose generality. In this way we will try, as it is our initial attention, to provide fluent passing through the text, and easy to understand abstract concepts. A reader will probably notice that we are repeating ourselves throughout the chapters. Nevertheless, it just seems so, and the real intention is to make things complicated gradually, and help a reader to go fluently through the text.

1.1 Complex system

Complexity emerges in many guises in the world around us, as well as in the world within us, and its abundance and influence on our lives impose a myriad of questions.[1] Even the research in complex systems is complex by itself, characterized by a plenty different approaches. Using the analogy of walking through the unknown field, we may say that even the guides are not so familiar with every part of the terrain, nevertheless that gives a beauty and the intricacy to the research field.

The essential characteristic of complex systems is the simplicity of its elements, on the one side, and noticeable irregularity which emerges from the collective interactions between elements, on the other side.

Probably the most essential attribute of complex system is that the interaction between its elements (or groups of elements) generates the appearance of new entity, with properties qualitatively different than the properties of its elements. This phenomenon is called the *emergence*. Although its meaning is a little blurry, it plays an important role in the distinguishing complex from simple systems side.

The existence of numerous definitions of complexity[2] and measures of complexity[3] is in the accordance with the juvenility of the research field on the one hand, and vagueness in the consensus of research community in accepting some of many properties of complex system as essential for fully characterizing it, on the other hand. Therefore, we will avoid this slippery field by starting the other way around. Namely, we will not start with some definition of complexity of a system, or eventually a measure of it, but we will start with some elementary properties that are, in more or less degree, present in variety of complex systems, in hope that together with many similar endeavors will participate in the collective evaluation of the essential properties of complex systems. Also our intention is to introduce an alternative method relied on the specific mathematical field for revealing emergent properties of some types of complex systems. Before we proceed with introducing general and widely accepted properties of real-world complex systems, let us take a look at some common examples.

The nature is full of the examples of complex systems. If we look to the sky we can notice a large group of birds, called a flock, flying together in a coherent formation, exhibiting synchronous behavior, and maneuvers they take look like they are orchestrated by some superior conductor. Nevertheless, it is quite opposite. They are flying in large groups from hundreds to thousands of them, and due to the local interactions between them a flock is moving and maneuvering without central control. In other words, the motion of a flock, that is spontaneous choice of a single common direction of flight, is a product of the behavior of lone animals, each one behaving exclusively on the basis on its own viewpoint. This collective animal behavior is not a rarity in nature, namely, bacteria, fish, insects, and herds exhibit similar behavior. Whereas the collective behavior of bird flock is mostly related to flying, the other animals like ants or termites perform different emergent collective actions. Let's take for example a termite mound. It is a rather complicated architectural construction built by a society of simple tiny animals, whose cooperative collective power exceeds that of the individual termite. Namely, the termite mound consist an array of bubble-like chambers which are connected by passages, and furthermore they are built as an energy efficient buildings controlling the flow of cold and hot air. If someone or something damages their mound, the whole colony of termites is awakened and groups with different roles take their actions, like soldiers gather for a battle and workers gather for rebuilding damaged part of mound. As in the case of bird flocks, this large super organism[4] exhibits complex behavior only from interactions of simple elements, i.e., termites.

What is common to these two exemplary systems, and many others, represents properties of complex systems in general, and can be summarized in the following. They exhibit self-organized (meaning without central control) collective behavior, which originates from local interactions between the elements of complex system, and new qualitatively different patterns (or behaviors or functions) emerge which cannot be recovered by knowing solely entities in complex system and their interactions. In the previous sentence we used the most frequent words in complex systems research (alongside with the notion of the "complexity" itself): *self-organization, collective behavior, no central control,* and *emergence.*[2] In the notions of these words are contained the essential properties of complex systems, and further the properties for distinguishability between complex, complicated, and simple systems, hence they need some clarifications. The self-organization is tightly related to the collective behavior and the absence of central control, meaning that the system formed by relatively simple elements interacting between themselves with simple interactions take actions as a group without guidance or control by some another entity. Take for example termites, and their taking an action toward the repair of a damaged mound. They are simple beings which display an interaction only with their immediate neighborhood, nevertheless, when mound is damaged, they spontaneously recruit and large groups of termites rush with mouth full of dirt toward the place of demolition. Whereas these notions are more or less related to the aggregations of entities, the notion of *emergence* is related to the function, or more precisely related to the manifestation revealed through that cooperative collective behavior. The change of formation, direction or shape for the flock of birds, are revealed through the change of different emergence patterns.

A word "system" is not restricted solely to the complex systems, but there are also simple systems and complicated systems. The latter one is particularly interesting since it can be incorrectly interpreted as a complex. A simple pendulum is an example of simple system, namely it is formed by the small number of elements, and the behavior of the system depends on well-known and understood laws (Newton's laws). Now, let's pay attention on a system with large number of elements, say few million, where each element is matched (or more generally, connected) with accurately suitable other elements. The whole collection of matched elements has to work in accordance so that the function of the system is performed. Furthermore, if some important element is damaged and out of order, the whole system may stop functioning. The difference with complex systems is rather obvious in all essential parts: in complex system each element does not have to have its appropriate match, the damage or removal of not just one but a fraction of elements does not necessarily affect the function of whole system. Removing a fraction of termites in termite colony and after that damaging the mound will not change the function of the rest of colony: they will self-organize to recruit all remaining forces to rebuild damaged mound. On the other hand take for example a watch. It has many parts, each part is perfectly matched with some other parts and they all work in unison, and as a result show us what time it is. The dysfunction of some key part does not make us too much harm, we take it to repairman and fix it. But there is another example of complicated system which has few million elements, and the dysfunction of one element or small fraction of elements, can make us harm. That system is an airplane. And further if we consider a captain as a part of the whole system, we can say that this system has central control. Although an airplane is not a complex system it is a part of large complex system, that is, an airline network of intertwine airports which originates from flight connections, and it is also a part of the network of the exchange of people and goods between cities and countries in which airports are located.

In practice there are two approaches to the research of complex systems, which are mutually complementary. In the first one researchers focus on the particular type, or more precisely a collection of data, of complex system (like ecological system, financial system, brain system, ...), and try to abstract general laws governing the behavior and function of those systems. On the other hand, there are practitioners whose research is focused on the developing theory, as well as tools, applicable on a broad sample of complex systems, regardless of their diverse functions. The content of this book, due to its generality, belongs to the latter approach.

Currently used mathematical tools have some drawbacks, on which we will point throughout the text, since they do not capture adequately higher-order relationships in complex systems, and accordingly we need to introduce new, or more precisely an alternative, mathematical framework for tackling problems which arise from these relationships.

We have introduced some basic notions of complex systems necessary for grasping into some particular subsets of this broad and expanding research field. From simple properties and very general rules, large groups of entities form remarkable endeavors to which we admire and from better understanding of those endeavors we can improve our, man-made systems. In introducing general notions and concepts we will try, as we did it so far, to illustrate it with appropriate examples whenever it is possible. The authors' intention is not to introduce mathematical concepts in a rigorous way, but to set the applicable framework for further development and tackling problems in the field of complex dynamical systems.

1.2   Complex networks

Perhaps the most important subfield in complex systems research is the research in complex networks, which is focused on a widespread of (sometimes unrelated) real world phenomena, and unifies researchers from different fields into a common task which exceeds the research in their native fields. Clearly, such endeavor is followed by the bursts in technology, mainly in the information technology, that is the fast growing development of the Internet, which provided an easy availability of data. And further, mathematical objects from graph theory defined as sets of elements and together with their pairwise relations have been found convenient for easy customization of a broad research community. The beginnings of graph theory originate in the work of Euler[5] and his famous Königsberg bridges problem. Namely, the city of Königsberg, now Kaliningrad in Russia, had seven bridges which connected four parts of the city, and Euler was puzzled whether someone can cross all seven bridges without crossing any bridge twice. He noticed that for solving that problem, the distances are irrelevant, but the connectivity is what actually matters, hence he presented different parts of the city as points and bridges connecting them as lines (or edges). In this particular case the answer was negative, that is, one cannot visit all four parts of the city by crossing every bridge only once. We will not go into the details of his solution, but the emphasis is on the way he derived it, that is, by noting that we can neglect distances and that it is enough to deal with points (or vertices or nodes) and connections between them (called edges or links). In other words, this simplification actually helped to solve a problem accurately, and established a birth of new mathematical field: graph theory.

From the example of Königsberg bridges problem we can notice that we know exactly whether two nodes are connected or not, and we know exactly how many edges each node has. But what happens when we have a large number of nodes and large number of links? This new situation brings us to two important persons in the research of graphs, and that is Paul Erdős and Alfred Rényi. Namely, among many important mathematical problems that they tackled, the one that is particularly interesting for us is the introduction of random graphs. Let us consider a graph with large number of nodes and large number of links, like online social network or the Internet (in the sense of remote connectivity between computers). A complete knowledge of connectivity between every pair of nodes is often lacking or unnecessary or varying over time, hence for the analysis of the large scale networks new tools and methods are necessary to devise. Erdős and Rényi developed the *random graph theory*[6, 7] following the simple rules for graph's construction: the graph represents the set of different vertices and the set of edges which connect randomly selected pairs, or in an alternative way of construction, edges are present with probability p. Interestingly, such graphs display the small-world property, which means that most pairs of vertices are connected by a short path through the graph, just like it is observed in the real world experiments.[8]

Nevertheless, there is another property observed in real world networks which is lacking in the random graphs, and that is high clustering,[9] which can be in plain words (in the context of social networks) defined as "the friend of your friend is likely also to be your friend", or in more general way, as the fraction of interactions between nearest neighbors of vertex among each other divided by the maximum number of possible connections between nearest neighbors. In order to satisfy both properties (small-worldness and high clustering) Watts and Strogatz designed the so called *small-world network model*[10] which, for some range of parameter displays both properties. It seemed that the problem of modeling real world networks is close to solution, however, there was another property which is not satisfied by either the random graph model or the small-world network model. Namely, in networks, either artificial or real, the vertex does not have to have the same number of neighbors, and the property which characterizes it is the distribution of connections per vertex. Hence, the calculation of probability distribution function $P(k)$ gives the probability that a randomly selected vertex has k neighbors.[11] In the case of random network the connections between vertices are placed randomly, and accordingly the majority of vertices have approximately the same number of neighbors, near the average number of connections of the network, satisfying the properties of the Poissonian-like behavior of the distribution function of connections. Although the networks build within the framework of the small-world model have two important properties, the distribution of connections is similar to that of the random network. On the other hand, for the most real-world networks it was found that the distribution of connections follows the power-law behavior.[12] In order to overcome that obstacle, Albert and Barabási proposed the *scale-free network model*,[13] which in its essence incorporates two mechanisms characteristic for many real-world networks: the growth and the preferential attachment. The first mechanism, the growth, simple means that networks are growing by adding new vertices and connecting them to already present vertices in the network, whereas the likelihood to connect to the vertex already present in the network depends on the number of neighbors the vertex has. In other words, the more connection the vertex has, the higher probability that the newcomer will connect to it. The outcome of application of these simple mechanisms in network formation is the network with the small-world property, high clustering, and the power-law distribution of connections.

This list of complex network models is not complete, nevertheless, we want to point out on the most important models which represent the milestones in the complex networks research. And the list of properties as well as phenomena, related to the complex networks is much more richer.[14, 15] For example, there is a large set of critical phenomena related to networks,[16] like the epidemic spreading, the percolation, the random and intentional failures taking place on networks, to name a few. On the other hand, the emergence of mesoscopic structures, like densely connected groups of vertices, called communities,[17] plays an important role in shaping the network's structure.

1.3 Simplicial complexes

Unlike the research in complex networks, the comprehensive overview of the variety of research in simplicial complex is still lacking. Since one of the purposes of this book is to promote versatile research applications of simplicial complexes, we will give a short review of the simplicial complexes applications in diverse research fields, hence this section of the Introduction will be different compared to the previous two sections.

Simplicial complexes as mathematical objects can be defined in different ways, as it will be introduced in this book, but probably the simplest definition which can help us to imagine simplicial complex is defining it as *a set of connected (glued along the common sides) polyhedra (generalized pyramides) which build a higher dimensional discrete geometrical space.* It turns out that the typical properties (like emergence) of complex systems can be easily captured by simplicial complexes.

The idea of modeling complex systems by analyzing its elements represented by simplices is not a new one. Namely, Ronald Atkin,[18, 19] following the ideas of Dowker[20] of building a simplicial complex from the relations between the elements of two sets (or the same set), have introduced the method of Q-analysis.[21] Researchers have sporadically used the methods of Q-analysis for the analysis of specific systems, often with small number of elements. Such cases span from studying qualitative and quantitative structure of television program,[22] analysis of the content of newspaper stories,[23] social networks,[24–27] urban planning,[28, 29] relationships among geological regions,[30] distribution systems,[31] decision making,[32] diagnosis of failure in large systems,[33] to mention a few. From this short overview of the applications we can see the wide range of systems to which Q-analysis can be applied. Recently Atkin's methodology received a further development in the work of Barcelo and Laubenbacher,[34] naming their theory an A-homotopy theory in honor of Atkin.

In modern theoretical physics simplicial complexes appear as important and convenient objects[35–37] due to their analytical and computational convenience. Namely, the language of modern physics is based on the calculus on manifolds which are discretized using simplicial complexes, and vice versa, simplicial complexes can be used for the study of topological properties of a manifold obtained from experimental data.[38] On the other hand, the use of simplicial complexes in discretization of exterior differential forms is extremely important. It is now widely recognized that geometry and topology are at the foundation of many physical theories such as general relativity,[39, 40] electromagnetism,[41] gauge theory,[42] elasticity,[43] to mention a few. For example, the development of simplicial quantum gravity[44] depends on the results of the Regge calculus,[45] which, in turn, was developed by approximating smooth 4-dimensional manifold by rigid simplices. On the other hand, in the computational electromagnetism[46–48] Maxwell's equations can be directly expressed in terms of discrete differential forms which are defined as cochains on simplicial complexes. The geometric and topological nature of such theories is often concealed by their formulation in vectorial and tensorial forms because of the unavoidable use of coordinate systems so that the complete topological and geometrical nature is obscured hiding for example, local and global invariants. Exterior derivative of differential forms is, on the other hand, invariant under a coordinate system change and since every differential equation may be expressed in terms of exterior derivative of differential forms,[49] many physical laws may be expressed in terms of differential forms. Discretization of differential forms using finite differences, for example, and using their coordinate values leads to numerical invalidation of some basic theorems (Stokes, for example) making traditional discretization methods pointless. It turns out that proper discretization of differential forms that preserves all the fundamental differential properties is possible only on simplicial complexes.[50] As a curiosity let us mention that there are some attempts to formulate physical theories in a completely discrete fashion,[51–54] with emphasis on simplicial complexes.

Due to the rapid achievements in informational technologies, the researchers, as well as many industries, are overwhelmed with large data sets waiting to be analyzed and treated in the right manner. The topological data analysis[55, 56] emerges as a collection of methods which originate from the algebraic topology and as mathematical objects use simplicial complexes. The goal of topological data analysis is to apply topology and to develop tools which may lead toward an insight to the shape of data, or in other words, the geometric properties of data. So far, the method have been proven successful in practice through the applications in different fields, and different data resources, like the neuroscience research,[57–59] sensor networks,[60, 61] computer vision,[62] population activity in visual cortex,[63] evolutionary tree,[64] protein classification,[65] protein folding,[66, 67] musical data,[68] sports analytics,[69] text mining,[70] dynamical systems,[71] to name a few applications. This list of applications is far from being complete, but even from this small sample it is easy to get an insight to the great potential of topological data analysis applications in different fields.

The concepts Q-analysis, as well as topological data analysis, are applied on the complex networks and dynamical systems[71–77] in the course of better understanding the structural relationships which emerge from network's mesoscopic structure. On the other hand, simplicial complexes turned to be useful in the social dynamics modeling representing suitable mathematical objects for modeling opinions.[78–80]

Considering the importance of simplicial complexes in the fundamental theories, we may consider simplicial complexes as universal tools for comprehensive and wide encompassing study of complex systems.

Bibliography

[1] MITCHELL M. Complexity: A guided tour [M]. Oxford: Oxford University Press, 2011.

[2] LADYMAN J, LAMBERT J, WIESNER K. What is a complex system? [J]. Euro. Jour. Phil. Sci., 2013, 3: 33.

[3] LLOYD S. Measures of complexity: A nonexhaustive list [J]. IEEE Control Systems Magazine, 2001, 21(4): 7.

[4] HÖLLDOBLER B, WILSON E O. The Superorganism: The beauty, elegance, and strangeness of insect societies [M]. New York: W. W. Norton & Company, 2008.

[5] EULER L. Solutio problematis ad geometriam situs pertinentis [J]. Comment. Acad. Sci. U. Petrop., 1736, 8: 128–140. Reprinted in Opera Omnia Series Prima, 1976, 7: 1–10.

[6] ERDŐS P, RÉNYI A. On random graphs [J]. Publicationes Mathematicae, 1959, 6: 290.

[7] ERDŐS P, RÉNYI A. On the evolution of random graphs [J]. Publications of Mathematical Institute of the Hungarian Academy of Sciences. 1960, 5: 17.

[8] MILGRAM S. The small world problem [J]. Psychology Today, 1967, 2: 60.

[9] ALBERT R, BARABÁSI A-L. Statistical mechanics of complex networks [J]. Rev. Mod. Phys., 2002, 74: 47.

[10] WATTS D J, STROGATZ S H. Collective dynamics of small-world networks [J]. Nature, 1998, 393: 440.

[11] BOCCALETTI S, LATORA V, MORENO Y, *et al.* Complex networks: Structure and dynamics [J]. Phys. Rep., 2006, 424: 175.

[12] CALDARELLI G. Scale-free networks: Complex webs in nature and technology [M]. Oxford: Oxford University Press, 2007.

[13] ALBERT R, BARABÁSI A-L, JEONG H. Mean-field theory for scale-free random networks [J]. Physica A, 1999, 272: 173.

[14] COHEN R, HAVLIN S. Complex networks: Structure, robustness and function [M]. Cambridge: Cambridge University Press, 2010.

[15] NEWMAN M. Networks: An introduction [M]. Oxford: Oxford University Press, 2010.

[16] DOROGOVTSEV S N, GOLTSEV A V, MENDES J F F. Critical phenomena in complex networks [J]. Rev. Mod. Phys., 2008, 80: 1275.

[17] FORTUNATO S. Community detection in graphs [J]. Phys. Rep., 2010, 486: 75.

[18] ATKIN R H. From cohomology in physics to q-connectivity in social sciences [J]. Int. J. Man-Machine Studies, 1972, 4: 341.

[19] ATKIN R H. Combinatorial connectivities in social systems [M]. Stuttgart: Birkhäuser Verlag, 1977.

[20] DOWKER C H. Homology groups of relations [J]. Annals of Mathematics, 1952, 56: 84.

[21] ATKIN R H. Mathematical structure in human affairs [M]. London: Heinemann, 1974.

[22] GOULD P, JOHNSON J, CHAPMAN G. The structure of television [M]. London: Pion Limited, 1984.

[23] JACOBSON T L, YAN W. Q-Analysis techniques for studying communication content [J]. Quality & Quantity, 1998, 32: 93–108.

[24] SEIDMAN S B. Rethinking backcloth and traffic: Prespectives from social network analysis and Q-analysis [J]. Environment and Planning B., 1983, 10: 439–456.

[25] FREEMAN L C. Q-analysis and the structure of friendship networks [J]. Int. J. Man-Machine Studies, 1980, 12: 367–378.

[26] DOREIAN P. Polyhedral dynamics and conflict mobilization in social networks [J]. Social Networks, 1981, 3: 107–116.

[27] DOREIAN P. Leveling coalitions as network phenomena [J]. Social Networks, 1982, 4: 27–45.

[28] ATKIN R H, JOHNSON J, MANCINI V. An analysis of urban structure using concepts of algebraic topology [J]. Urban Studies, 1971, 8: 221–242.

[29] JOHNSON J H. The Q-analysis of road intersections [J]. Int. J. Man-Machine Studies, 1976, 8: 531–548.

[30] GRIFFITHS J C. Geological similarity by Q-analysis [J]. Mathematical Geology, 1983, 15: 85.

[31] DUCKSTEIN L. Evaluation of the performance of a distribution system by Q-analysis [J]. Applied Mathematics and Computation, 1983, 13: 173–185.

[32] DUCKSTEIN L, NOBE S A. Q-analysis for modeling and decision making [J]. European Journal of Operational Research, 1997, 103: 411–425.

[33] ISHIDA Y, ADACHI N, TOKUMARU H. Topological approach to failure diagnosis of large-scale systems [J]. IEEE Trans. Syst., Man and Cybernetics, 1985, 15: 327–333.

[34] BARCELO H, LAUBENBACHER R. Perspectives on a-homotopy theory and its applications [J]. Discrete Mathematics, 2005, 298: 39.

[35] FLANDERS H. Differential forms with applications to the physical sciences [M]. New York: Academic Press, 1963.

[36] FRANKEL T. The geometry of physics: An introduction [M]. Cambridge: Cambridge University Press, 1997.

[37] ESCHRIG H. Topology and geometry for physics [M]. Heidelberg: Springer-Verlag, 2011.

[38] MULDOON M R, MACKAY R S, HUKE J P, *et al.* Topology from time series [J]. Physica D., 1993, 65: 1.

[39] MISNER C W, THRONE K S, WHEELER J A. Gravitation [M]. San Francisco: W.H. Freeman, 1973.

[40] FRAUENDIENER J. Discrete differential forms in general relativity [J]. Classical and Quantum Gravity, 2006, 23(16): S369–S385.

[41] BOSSAVIT A. Computational electromagnetism: Variational formulations, complementarity, edge elements [M]. New York City: Academic Press, 1998.

[42] CHRISTIANSEN S H, HALVORSEN T G. A simplicial gauge theory [J]. J. Math. Phys, 2012, 53: 033501.

[43] YAVARI A. On geometric discretization of elasticity [J]. Journal of Mathematical Physics, 2008, 49(2): 1–36.

[44] HUMBER H W. Simplicial quantum gravity, in: Critical phenomena, random systems, gauge theories [M]. Amsterdam: North-Holland, 1986.

[45] REGGE T. General relativity without coordinates [J]. Nuovo Cimento, 1961, 19: 558

[46] DESCHAMPS G A. Electromagnetics and differential forms [J]. Proceedings of IEEE, 1981, 69: 676.

[47] TEIXEIRA F L. Geometric aspects of the simplicial discretization of Maxwell equations [J]. Progress in Electromagnetics Research, 2001, 32: 171–188.

[48] TONTI E. Finite formulation of electromagnetic field [J]. IEEE Trans. Mag., 2002, 38: 333.

[49] SHARPE R W. Differential geometry: Cartan's generalization of klein's erlangen program [M]. New York: Springer-Verlag, 1997.

[50] GAWLIK E, MULLEN P, PAVLOV D, *et al.* Geometric, variational discretization of continuum theories [J]. Physica D., 2011, 240: 1724.

[51] ATKIN R H. Abstract physics [J]. Nuovo cimento, 1965, 38: 496.

[52] ATKIN R H, BASTIN T. A homological foundation for scale problems in physics [J]. International Journal of Theoretical Physics, 1970, 3: 449.

[53] TONTI E. The reason for analogies between physical theories [J]. Appl. Math. Modelling, 1976, 1: 37.

[54] TONTI E. A direct discrete formulation of field laws: The cell method [J]. Comput. Model. Eng. Sci., 2001, 2: 237.

[55] CARLSSON G. Topology and data [J]. American Mathematical Society Bulletin, 2009, 46(2): 255.

[56] EPSTEIN C, CARLSSON G , EDELSBRUNNER H. Topological data analysis [J]. Inverse Problems, 2011, 27(12): 120201.

[57] DABAGHIAN Y, MÉMOLI F, FRANK L, *et al.* A topological paradigm for hippocampal spatial map formation using persistent homology [J]. PLoS Comput. Biol., 2012, 8(8): e1002581.

[58] CHUNG M K, HANSON J L, YE J, *et al.* Persistent homology in sparse regression and its application to brain morphometry [J]. IEEE T M., 2015, 34: 1928.

[59] BENDICH P, MARRON J S, MILLER E, *et al.* Persistent homology analysis of brain artery trees [J]. Ann. Appl. Stat., 2016, 10: 198.

[60] DE SILVA V, GHRIST R. Coordinate-free coverage in sensor networks with controlled boundaries via homology [J]. The International Journal of Robotics Research, 2006, 25: 1205.

[61] DE SILVA V, GHRIST R. Coverage in sensor networks via persistent homology [J]. Algebraic & Geometric Topology, 2007, 7: 339.

[62] CARLSSON G, ISHKHANOV T, DE SILVA V, *et al.* On the local behavior of spaces of natural images [J]. Int. J. Comput. Vis., 2008, 76: 1.

[63] SINGH G, MEMOLI F, ISHKHANOV T, *et al.* Topological analysis of population activity in visual cortex [J]. Journal of Vision, 2008, 8: 11.

[64] CHAN J M, CARLSSON G, RABADAN R. Topology of viral evolution [J]. PNAS., 2013, 110: 18566.

[65] CANG Z, MU L, WU K, *et al.* A topological approach for protein classification [J]. arXiv:1510.00953.

[66] KRISHNAMOORTHY B, PROVAN S, TROPSHA A. A topological characterization of protein structure [J]. Data Mining in Biomedicine, 2007, 7: 431.

[67] XIA K, WEI G W. Persistent homology analysis of protein structure, flexibility, and folding [J]. Int. J. Numer. Method. Biomed. Eng., 2014, 30: 814.

[68] SETHARES W A, BUDNEY R. Topology of musical data [J]. Journal of Mathematics and Music: Mathematical and Computational Approaches to Music Theory, Analysis, Composition and Performance, 2014, 8: 73.

[69] GOLDFARB D. An application of topological data analysis to hockey analytics [J]. arXiv:1409.7635.

[70] WAGNER H, DŁOTKO P, MROZEK M. Computational topology in text mining [C]// FERRI M, FROSINI P, LANDI C, *et al.* Computational Topology in Image Context. Bertinoro, Italy: CTIC, 2012: 68–78.

[71] MALETIĆ S, ZHAO Y, RAJKOVIĆ M. Persistent topological features of dynamical systems [J]. Chaos., 2016, 26: 053105.

[72] MALETIĆ S, RAJKOVIĆ M, VASILJEVIĆ D. Simplicial complexes of networks and their statistical properties [J]. Lecture Notes in Computational Science, 2008, 5102(II): 568–575.

[73] MALETIĆ S, STAMENIĆ L, RAJKOVIĆ M. Statistical mechanics of simplicial complexes [J]. Atti Semin. Mat. Fis.Univ. Modena Reggio Emilia, 2011, 58: 245–261.

[74] MALETIĆ S, RAJKOVIĆ M. Combinatorial Laplacian and entropy of simplicial complexes associated with complex networks [J]. Eur. Phys. J. Special Topics, 2012, 212: 77.

[75] MALETIĆ S, HORAK D, RAJKOVIĆ M. Cooperation, conflict and higher-order structures of complex networks [J]. Advances in Complex Systems, 2012, 15: 1250055.

[76] ANDELKOVIĆ M, TADIĆ B, MALETIĆ S, *et al.* Hierarchical sequencing of online social graphs [J]. Physica A, 2015, 436: 582.

[77] HORAK D, MALETIĆ S, RAJKOVIĆ M. Persistent homology of complex networks [J]. J. of Stat. Mech., 2009, 03: P03034.

[78] MALETIĆ S, RAJKOVIĆ M. Simplicial complex of opinions on scale-free networks [J]. Studies in Computational Intelligence, Springer, 2009, 207: 127–134.

[79] MALETIĆ S, RAJKOVIĆ M. Consensus formation on simplicial complex of opinions [J]. Physica A., 2014, 397: 111–120.

[80] MALETIĆ S, ZHAO Y. Hidden multidimensional social structure modeling applied to biased social perception [J]. Physica A, 2018, 492: 1419–1430.

The World of Simplicial Complexes

So far we have seen that the richness of complexity phenomena neces-
sitates the richness of tools for tackling the problems which emerge
from real-world phenomena, and as well the widespread application
of mathematical formalisms which rely on the simplicial complexes.
In this chapter we will give an introduction to simplicial complexes,
which includes defining them, definition of their properties, and quantities which characterize them. As we mentioned earlier, whenever its
convenient we will avoid formal mathematical definition, and reach
out to the intuitive graphical or real world examples. Although the
way of defining simplices is a bit unusual, our intention is to point
the equivalence of three definitions, or in other words, that the same
mathematical object can be defined in three different ways. The versatility of descriptions of simplicial complex originates from three
distinct aspects from which we can treat them: combinatorial, geometrical, and topological. Therefore, we hope that by the end of this
chapter the reader will have at least intuitive image which will help
for easy flowing through the forthcoming chapters.

2.1 Geometric simplicial complex

Our introducing of simplicial complexes will start with a definition of geometric simplicial complex,[1] although due to the equivalence between definitions the order is not important, we have chosen
to introduce a geometrical simplicial complex because of practical

reasons. Namely, it will provide us a background for graphical representation of simplices and their relations, and hence make easier understanding of becoming abstract definitions.

Let us take a two line segments (called edges) in some space, say $\mathbf{R}^N$, with one end the same. We can glue them along a common end, called vertex. Next we can take two triangles in the same space, which have a common edge, and we can glue them along that common edge. We can further take two pyramid-like tetrahedra with a common side (triangle), and we can glue along the common side. By continuing in the same way, we can do the same with higher dimensional analogues by gluing them along the common sides (called faces). Clearly, by aggregating these "building blocks" in space we are building a polyhedral structure in space.

After this intuitive picture, let's turn toward more formal definitions. Take a set of points $V = \{v_0, v_1, v_2, \ldots, v_m\}$ in an Euclidean space $\mathbf{R}^N$, and let for any $c_i \in \mathbf{R}$, set V satisfies the equations

$$\sum_{i=0}^{m} c_i v_i = 0 \quad \text{and} \quad \sum_{i=0}^{m} c_i = 0$$

which imply that $c_1 = c_2 = \cdots = c_m = 0$, then we say that set V is geometrically independent. It is easy to show that V is geometrically independent if and only if the vector entries

$$v_1 - v_0, \cdots, v_m - v_0$$

are linearly independent. After defining geometrically independent set, we are ready to define an n-plane spanned by these points, such that the n-plane consists all points x of $\mathbf{R}^N$ such that

$$x = \sum_{i=0}^{m} c_i v_i = 0 \quad \text{where} \quad \sum c_i = 1 \tag{2.1}$$

On the other hand, an n-plane P can be also defined as a set of all points x such that

$$x = v_0 + \sum_{i=1}^{m} c_i (v_i - v_0)$$

for some scalars $c_1, \ldots, c_m$. This expression means that plane P is actually a plane through v_0 parallel to the vectors formed by $v_i - v_0$. After last definition it is easy now to define the n-simplex σ to be the set of all points x in $\mathbf{R}^N$ such that the condition (2.1) is satisfied, and $c_i \leq 1$ for all i. The coefficients c_i are called "the barycentric coordinates of the point x of simplex σ with respect to $c_1, \ldots, c_m$".[1] In this sense, 0-simplex is a point, called also a vertex. The 1-simplex is a line segment, called an edge, and it is spanned by vertices v_0 and v_1, and contains points

$$x = ca_0 + (1 - c)a_1$$

Spanning the 2-simplex, i.e., the triangle, with v_0, v_1, v_2 as vertices, goes as follows. Take again all

$$x = \sum_{i=0}^{2} c_i v_i = c_0 v_0 + (1 - t_0) \left[\frac{c_1}{\mu} v_1 + \frac{c_2}{\mu} v_2 \right]$$

where $\mu = 1 - t_0$. In Fig. 2.1 is indicated point p on an edge $v_1 v_2$ expressed by $\left[\frac{c_1}{\mu} v_1 + \frac{c_2}{\mu} v_2 \right]$. Hence, point x is a point on the line between v_0 and p. In this sense, we can understand building a 2-simplex as "sliding" all lines between point v_0 and all points on a line segment between and including v_1 and v_2.

From these examples, as well from the definition, an n-simplex σ_n is spanned over $n + 1$ points, and we say that n is dimension of simplex σ_n. It is easy to generalize this way of building simplices with dimensions higher than 2, resulting in tetrahedra, pentahedra, and in general-polyhedra[2] (Fig. 2.2).

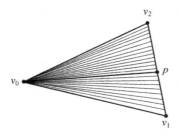

Fig. 2.1　An example of building a 2-simplex by sliding lines

Simplex	Common name	Geometric realisation
0-simplex	point vertex node	
1-simplex	line edge link	
2-simplex	triangle	
3-simplex	tetrahedron	
4-simplex	pentahedron	

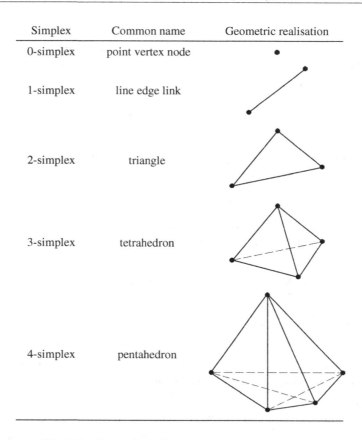

Fig. 2.2   Examples of higher-dimensional simplices

Any subset of spanning points of a simplex that preserves geometrical independency is also a simplex and it is called a face of simplex.[2] Fig. 2.3 illustrates a 3-simplex together with its 0-simplex (v_0), and 2-simplex $([v_1, v_2, v_3])$, although 3-simplex contains in total four-simplices, six 1-simplices and four 2-simplices.

Finally, a geometric simplicial complex K is a finite set of simplices in some Euclidean space $\mathbf{R}^N$ such that[1]:

(i) if simplex σ_n is in K, and τ_p is a face of σ_n, then τ_p is in K; and

(ii) if two simplices σ_n and τ_r are in K, then $\sigma_n \cap \tau_r$ is either empty, or is a common face of σ_n, then τ_p.

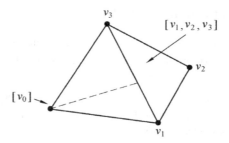

Fig. 2.3   Examples of different dimensional faces of 3-simplex

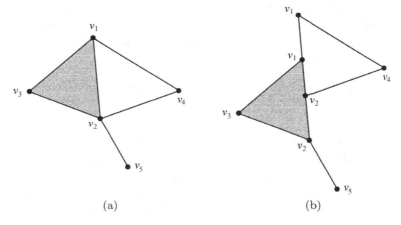

(a)   (b)

Fig. 2.4   An example of geometrical realizations of different dimensional simplices as a simplicial complex (a), and a collection of simplices which does not build a simplicial complex (b)

The dimension of simplicial complex K $(\dim(K))$ is the maximum dimension of all simplices.

Defined in this way a simplicial complex K is not topological space by itself, but the set of points of $\mathbf{R}^N$ that lie in simplices of K are actually topologizing a subspace of $\mathbf{R}^N$, and thus forming a polyhedron. An example of geometrical realizations of different dimensional simplices as a simplicial complex is illustrated in Fig. 2.4(a), whereas collection geometrical simplices in Fig. 2.4(b) is not a simplicial complex.

2.2 Abstract simplicial complex

We have finished previous section by defining the geometrical simplicial complex. Now, recall that we started our path toward this definition by taking some set of points $V = \{v_1, v_2, \ldots, v_m\}$ in Euclidean space. If we assume for a moment that geometrical positions of points are well defined, from geometrical definition of a simplex, any simplex can be characterized only as a sequence of points, i.e., vertices, meaning that an n-simplex is a subset of $n+1$ elements of the set V. Hence, a set of simplices and their faces is actually a subset of power set of V, or in other words, subset of the set of subsets of V.

Let us start again with a set $V = \{v_1, v_2, \ldots, v_m\}$, and let $P(V)$ be power set (set of all subsets) of V. A family of subsets of $P(V)$, labeled by K, is called an *abstract simplicial complex*[2] if it is closed under taking subsets, i.e., the following is satisfied

(i) if $\sigma \in K$ and τ is subset of σ, then $\tau \in K$, and
(ii) for all $\{v\} \in V$ is also $\{v\} \in K$.

An element of set V is called a vertex, whereas an element of the set K is called simplex. Subset τ of σ is called a face of σ, denoted by $\tau \leq \sigma$. Similarly to the definition of simplices in the geometrical case, the dimension of a simplex σ is equal to its cardinality minus one, that is, the number of elements of an element σ from K minus one.

In order to make this definition more understandable, we will illustrate it by a simple example. The vertex set is $V = \{v_1, v_2, v_3, v_4, v_5\}$ and a family of subsets, that is simplices

$$K = \left\{ \begin{array}{c} \{v_1\}, \{v_2\}, \{v_3\}, \{v_4\}, \{v_5\}, \{v_1, v_2\}, \{v_1, v_3\}, \{v_1, v_4\}, \\ \{v_2, v_3\}, \{v_2, v_4\}, \{v_2, v_5\}, \{v_1, v_2, v_3\} \end{array} \right\}$$

If we compare this set of abstract simplices with a geometrical simplicial complex in Fig. 2.4(a) we can notice that the obvious similarity, that is both structures (geometrical and combinatorial) represent the same relationships between elements. With a simple inspection we see that simplicial complex K is built by five 0-simplices, six 1-simplices, and one 2-simplex; 1-simplices $\langle v_1, v_2 \rangle$, $\langle v_1, v_3 \rangle$, $\langle v_2, v_4 \rangle$ are 1-faces of 2-simplex $\langle v_1, v_2, v_3 \rangle$.

2.3 Simplicial complex via relation

In the previous section we saw that by taking collections of elements (i.e., subsets) of some set we can build an abstract simplicial complex. Also, recalling the example at the end of previous section we see that not all subsets of a set are taken in building a simplicial complex. Now we can wonder whether we can impose some criteria for taking these collections of elements (i.e., subsets), or whether mathematical formalism of algebraic topology allows us impose some criteria. Thanks to the work of Dowker,[3] and further extensively developed by Atkin,[4–6] introducing a criteria for making simplices results in generating two simplicial complexes, which makes the tools of algebraic topology even more powerful in the applications to real-world phenomena.

But before introducing yet another abstract definition we will consider some examples which may give us a hint for this new definition. Suppose that we have two sets, say A and B, and suppose that the elements of set A are by some rule related to the elements of set B. For example, the elements from the set A may correspond to the individuals and the elements from the set B may correspond to the diverse interests of each individual and the relation may correspond to the property "person from set A has an interest in set B". As another example, we may assume that the elements from the set A correspond to patients, where as elements from the set B correspond to diverse clinical symptoms, and the relation corresponds to the property "patient from set A has a symptom from set B". Or, the elements from the set A may correspond to the city streets where as the elements from the set B may correspond to the diverse junctions and the relation may correspond to the property "street from set A contains a junction in set B". As another example consider that elements from the set A correspond to the TV shows whereas the elements from the set B correspond to the diverse subjects, covered by the show and the relation corresponds to the property "TV show from set A has a subject from set B". In the context of social issues the elements from the set A may correspond to the social groups, the elements from the set B may correspond to the diverse persons, and the relation may correspond to the property "social group from set A has as a member person from set B". As a final example, the elements from the set A may correspond to the geological regions, whereas the elements from the set B may correspond to the diverse rock types, and the relation may correspond to the property "geological region from set A has a rock type from set B", and so on. A careful reader probably noticed that we already mentioned these examples in the Introduction.

Let us now make the above real situations formalized. Alongside with the previously introduced set $V = \{v_1, v_2, \ldots, v_m\}$, we will introduced another set $S = \{s_1, s_2, \ldots, s_n\}$. A binary relation λ represents some rule or property which assigns to every element in S *one or more* elements in V, i.e., for every $s_i \in S$ exists $v_j \in V$ such that $s_i \lambda v_j$. The set S and the relation λ determine the subset K of the power set of V and we label each element $\{v_{\alpha_0}, v_{\alpha_1}, \ldots, v_{\alpha_q}\} \in K (q \leq m)$ by the element $s_i \in S$ for which $s_i \lambda v_{\alpha_0}, s_i \lambda v_{\alpha_1}, \ldots, s_i \lambda v_{\alpha_q}$. To distinguish the element s_i from the set S and its associated element from the set K due to the relation λ, the element of the set K will be labeled as $\sigma(s_i)$. Therefore, the notation $\sigma_q(s_i) = \{v_{\alpha_0}, v_{\alpha_1}, \ldots, v_{\alpha_q}\}$ [7–9] means that an element s_i of the set S is λ-related to q elements $\{v_{\alpha_0}, v_{\alpha_1}, \ldots, v_{\alpha_q}\}$ of the set V. The elements of the set V are called *vertices*, whereas the elements of the set K are called q-dimensional simplices or just q-simplices. Further, an element s_i is λ-related to any subset of the set $\{v_{\alpha_0}, v_{\alpha_1}, \ldots, v_{\alpha_q}\}$, and hence, every subset of $\{v_{\alpha_0}, v_{\alpha_1}, \ldots, v_{\alpha_q}\}$ is also a simplex, meaning that any such subset is a face of simplex, due to the definition of q-face. Since each $s_i \in S$ identifies a q-simplex $\sigma_q(s_i)$ (for some q) together with all its faces, this collection of simplices is called a simplicial complex K, which we will denote $K_s(V, \lambda)$.[5] Comparing this definition and the one in Sec. 2.2 it is obvious that in both cases sets K of subsets of V are the same, and the only difference is that the definition through the relation assigns a label (or name) from set S to simplices in K. Two important notes emerge at this moment.

Namely, if we in some way obtain a simplicial complex defined in Sec. 2.2, we can easily introduce a set S and recover (if it's possible) the rule (i.e., the relation) responsible for making a simplicial complex. The second note is also related to the sets. Namely, the relation that is responsible for building a simplicial complex as it is defined in this section can be between the elements of the same set, that is $V = S$.

Like in previous sections, we will illustrate our exposition with an example, and the reader may already assume which one. Let us suppose that we have two sets $V = \{v_1, v_2, v_3, v_4, v_5\}$ and $S = \{s_1, s_2, s_3, s_4\}$, and the relation λ such that the following simplices are formed:

$$\sigma_2(s_1) = \langle v_1, v_2, v_3 \rangle$$

$$\sigma_1(s_2) = \langle v_2, v_4 \rangle$$

$$\sigma_1(s_3) = \langle v_1, v_4 \rangle$$

$$\sigma_1(s_4) = \langle v_2, v_5 \rangle$$

and having in mind that every subsimplex (that is a face) is also a simplex, and comparing with the example in the section 2.2, we can see that it is the same simplicial complex, or in the geometrical representation as in Fig. 2.5(a).

Although two definitions and approaches carry similarity by taking the subsets of a set of vertices, considering simplicial complexes in

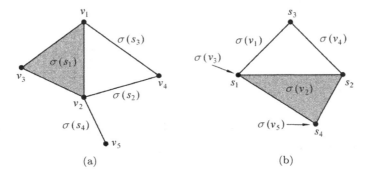

(a)   (b)

Fig. 2.5 (a) Example of simplicial complex, and (b) its conjugate

a way it is presented in this section may be more convenient for tackling practical problems. For example, consider a simplicial complex of streets (simplices) and junctions (vertices) in an urban area. Then from the above example, street s_1 contains junctions v_1, v_2, and v_3, whereas street s_2 contains junctions v_2, and v_4, and these two streets share a junction v_2. Hence, it is easy to comprehend how simplicial complex from Fig. 2.5(a) can capture the complicated relationships between streets which emerge through common junctions. Obviously, if we do not assign a street name with the corresponding junction, we would lose important information.

Since we are dealing with two sets, it gives us a certain arbitrariness in the choice of, say, vertex set, since we did not give any criteria for choosing a set V as vertex set. Maybe this sounds confusing, but in the example of streets and junctions we could say as well that at the junction v_1 streets s_1 and s_3 are crossing, or at the junction v_2 streets s_1, s_2, and s_3 are crossing, and so on. It is easy to anticipate now our intention. Namely, since the relation λ relates the elements of the set S with elements of the set V, there must be some relation which does the opposite, i.e., relates the elements of the set V with the elements of the set S. That role is taken by the inverse relation λ^{-1}[6, 7] of λ which relates the elements of the set V with the elements of the set $S : v_1\lambda^{-1}s_1, v_1\lambda^{-1}s_3, v_2\lambda^{-1}s_1, v_2\lambda^{-1}s_2$, and so on. Following the same procedure, we form a simplicial complex $K_V(S, \lambda^{-1})$ on the vertex set S defined by the relation λ^{-1}, illustrated in Fig. 2.5(b):

$$\sigma_1(v_1) = \langle s_1, s_3 \rangle$$

$$\sigma_2(v_2) = \langle s_1, s_2, s_4 \rangle$$

$$\sigma_0(v_3) = \langle s_1 \rangle$$

$$\sigma_1(v_4) = \langle s_2, s_3 \rangle$$

$$\sigma_0(v_5) = \langle s_4 \rangle .$$

Note that the elements of sets V and S have changed their roles, and in the complex $K_V(S, \lambda^{-1})$ simplices are from the set V whereas the vertices are from the set S. Now we can generalize these results to the simplicial complex defined by two arbitrary sets $S = \{s_1, s_2, \ldots, s_n\}$ and $V = \{v_1, v_2, \ldots, v_m\}$ and relation λ.

The n simplicial complex $K_V(S, \lambda^{-1})$ defined on the sets $S = \{s_1, s_2, \ldots, s_n\}$ and $V = \{v_1, v_2, \ldots, v_m\}$ by the inverse relation λ^{-1} of the relation λ is called *the conjugate complex* of the simplicial complex $K_S(V, \lambda)$.[6, 7] In order to clarify the importance of the simplicial complex and its conjugate, let us consider an example where the elements of the set S are patients, and the elements of the set V are clinical symptoms. Then the simplicial complex represents a collection of patients sharing the symptoms, whereas its conjugate complex represents a collection of clinical symptoms sharing the patients which have them.

Finally, as we already emphasized, the simplicial complex can be created on a single set, that is, following the above notation $S = V$ (that is $K_S(S, \lambda)$), hence the simplicial complex and its conjugate complex are the same.

2.4 Characterization of simplicial complexes

So far we have introduced three equivalent definitions of the same mathematical object, i.e., simplicial complex: geometrical, combinatorial, and by relation. Each of them has its advantages and drawbacks depending on the data which we want to represent as a simplicial complex and on a particular problem that we want to tackle. For convenience we will represent examples as a geometrical simplicial complex, regardless on the way how it was initially built. This will certainly help the reader to better understand upcoming definitions and concepts, on the one hand, and to serve to the interested reader to verify our results.

As in hitherto exposition we will use a single convenient example which will serve for illustration of tools and definitions. In Fig. 2.6 is illustrated a simplicial complex built from two sets S and V, represented by the letters and numerals, respectively.

In next two subsections we will introduce concepts and quanties which serve for the analysis of simplicial complexes, as well as for their distinguishing. Although the order of subsections does not follow the historical order, we will first introduce more intuitive and tangible concepts (Q-analysis), and then introduce more abstract (Homology).

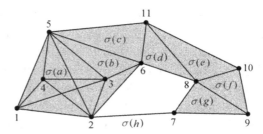

Fig. 2.6   Simplicial complex built from two sets, one represented by letters and the other by numerals

2.4.1   *Q-analysis*

In this subsection we will introduce a method, called Q-analysis, for the simplicial complexes analysis, introduced by R. Atkin[5, 10] in order to make social research a hard science,[4] like in the case of physics.

Incidence matrix

The relation between two sets S and V can be represented in matrix form, where the rows are associated with simplices and columns are associated with vertices. This matrix is called an *incidence matrix*. Λ, and the matrix entry $[\Lambda_{ij}]$ is equal to 1 if simplex $\sigma(i)$ contains a vertex j, and otherwise it is equal to 0. Hence, for the example from the Fig. 2.6 rows correspond to the elements of the set S, columns correspond to the elements of the set V, and matrix element $[\Lambda]_{ij}$ is equal to 1 if an element $s_i \in S$ is λ-related to the element $v_j \in V$

$$
\Lambda = \begin{bmatrix}
1 & 1 & 1 & 1 & 1 & 0 & 0 & 0 & 0 & 0 & 0 \\
0 & 1 & 1 & 0 & 1 & 1 & 0 & 0 & 0 & 0 & 0 \\
0 & 0 & 0 & 0 & 1 & 1 & 0 & 0 & 0 & 0 & 1 \\
0 & 0 & 0 & 0 & 0 & 1 & 0 & 1 & 0 & 0 & 1 \\
0 & 0 & 0 & 0 & 0 & 0 & 0 & 1 & 0 & 1 & 1 \\
0 & 0 & 0 & 0 & 0 & 0 & 0 & 1 & 1 & 1 & 0 \\
0 & 0 & 0 & 0 & 0 & 0 & 1 & 1 & 1 & 0 & 0 \\
0 & 1 & 0 & 0 & 0 & 0 & 1 & 0 & 0 & 0 & 0
\end{bmatrix}
$$

where rows are associated to simplices labeled by letters $a \sim h$, and columns are associated to vertices labeled by numerals $1 \sim 11$. It is easy to check that the incidence matrix of conjugate complex is just a transpose of the incidence matrix of the initial simplicial complex.

The matrix that captures the relationships between simplices is the so called *connectivity matrix*[7] defined as:

$$\Pi = \Lambda \cdot \Lambda^T - \Omega$$

where Λ is the incidence matrix, Ω is the matrix with all entries equal to 1, and "T" is the transpose operation on matrices. Rows and columns of the matrix Π are associated to simplices, the diagonal elements represent dimensions of simplices, whereas the off-diagonal elements represent the dimensionality of faces which simplices share. By convention, the entry $[\Pi_{ij}]$ means that two simplices do not share a face. For the example of simplicial complex from Fig. 2.6, the connectivity matrix has the following form:

$$\Pi = \begin{bmatrix}
4 & 2 & 0 & -1 & -1 & -1 & -1 & 0 \\
2 & 3 & 1 & 0 & -1 & -1 & -1 & 0 \\
0 & 1 & 2 & 1 & 0 & -1 & -1 & -1 \\
-1 & 0 & 1 & 2 & 1 & 0 & 0 & -1 \\
-1 & -1 & 0 & 1 & 2 & 1 & 0 & -1 \\
-1 & -1 & -1 & 0 & 1 & 2 & 1 & -1 \\
-1 & -1 & -1 & 0 & 0 & 1 & 2 & 0 \\
0 & 0 & -1 & -1 & -1 & -1 & 0 & 1
\end{bmatrix}$$

where rows and columns are associated to simplices labeled by letters $a \sim h$.

Q-vector

So far we have introduced the dimension of the simplex and the relationship (or adjacency) between two simplices through the shared common face, which are stored in the entries of connectivity matrix. Now we will introduce a higher aggregations of simplices induced through the shared face and, further, how they induce the intrinsic

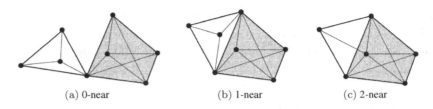

(a) 0-near   (b) 1-near   (c) 2-near

Fig. 2.7   An example of q-nearness between simplices

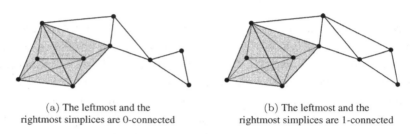

(a) The leftmost and the   (b) The leftmost and the
rightmost simplices are 0-connected   rightmost simplices are 1-connected

Fig. 2.8   An example of q-connectedness

hierarchical multilevel and multidimensional organization of simplicial complex. The property that any subsimplex of a simplex is also a simplex induces various levels of adjacency between simplices, and also various levels of connectivity between collections of simplices. Two simplices are *q-near* if they share a q-dimensional face (Fig. 2.7), and hence, they are also $(q-1)-, (q-2)-, \ldots, 1-$ and 0-near.

The collection of simplices in which any pair of simplices is connected by a sequence of simplices where a pair of successive simplices is q-near is called the *q-connected component*. More formally, two simplices σ and ρ are *q-connected*[4] if there is a sequence of simplices $\sigma, \sigma(1), \sigma(2), \ldots, \sigma(n), \rho$ such that any two consecutive simplices share at least a q-face. As an example of q-connectivity see Fig. 2.8. Note that if two simplices σ_p and σ_r are q-connected, they are also $(q-1)-, (q-2)-, \ldots, 1-,$ 0-connected in K.

The q-connectivity between simplices induces an equivalence relation on simplices of a complex K, since it is reflexive, symmetric, and transitive. This equivalence relation will be denoted by γ so that $(\sigma(i), \sigma(j)) \in \gamma_q$ if and only if $\sigma(i)$ is q-connected to $\sigma(j)$

Let K_q be the set of simplices in K with dimension greater than or equal to q, i.e., simplicial subcomplex built by simplices which have dimension greater than or equal to q. Then γ_q partitions K_q into equivalence classes of q-connected simplices. These equivalence classes are members of the quotient set K_q/γ_q and they are called the *q-connected components* of K. Every simplex in a q-component is q-connected to every other simplex in that component, but no simplex in one q-component is q-connected to any simplex on a distinct q-connected component. The cardinality of K_q/γ_q is denoted Q_q and is the number of distinct q-connected components in K. The value Q_q is the q-th entry of the so-called *Q-vector*[5, 11] (or *first structure vector*[7]), an integer vector with the length $\dim(K) + 1$. The values of the Q-vector entries are usually written starting from the number of connected components for the largest dimension in descending order, i.e.:

$$Q = [Q_{\dim(K)}Q_{\dim(K)-1}\cdots Q_1 Q_0]$$

An example illustrating the partitioning of the simplicial complex into q-connectivity classes and Q-vector for simplicial complex in Fig. 2.6 is presented in Fig. 2.9, with Q-vector entries:

$$Q = [1 \quad 2 \quad 6 \quad 2 \quad 1]$$

With first structure vector is closely related the so called *obstruction vector*.[7] If I is a unit vector of length $\dim(K)+1$ with all entries equal to 1, then the obstruction vector Q^* is defined as

$$Q^* = Q - I \text{ that is } Q = [Q_{\dim(K)} - 1 \quad Q_{\dim(K)-1} - 1 \quad \cdots \quad Q_0 - 1]$$

The obstruction vector quantifies the number of structural restrictions or obstructions to changes at q-levels, or in other words it enumerates the number of gaps on q-levels. For the working example from Fig. 2.6 the obstruction vector has the form

$$Q^* = [0 \quad 1 \quad 5 \quad 1 \quad 0]$$

It is of the essential importance to comprehend the meaning of first structure vector, and as well the obstruction vector. If we imagine that we are equipped with a special lenses with which we can "see" the structure of simplicial complex, then changing lenses we can "see" graded substructures of simplicial complex revealed through q-connectivity components.

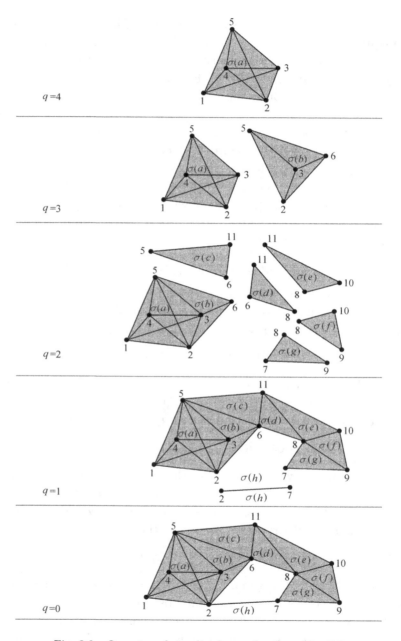

Fig. 2.9   Q-vector of simplicial complex from Fig. 2.6

Second structure vector

The second structure vector[7] n_q is an integer vector of length $\dim(K) + 1$. The value of q-th entry of this vector is equal to the number of simplices of dimension equal or larger than q, i.e., it is equal to the number of simplices at q-level of connectivity, with the notation:

$$n = [n_{\dim(K)} n_{\dim(K)-1} \cdots n_1 \ n_0]$$

Hence, for our working example, it has the form:

$$n = [1 \quad 2 \quad 7 \quad 8 \quad 8]$$

Third structure vector

The entries of the third structure vector[12] $\bar{Q}_q$ are defined in the following way:

$$\bar{Q}_q = 1 - \frac{Q_q}{n_q}$$

where Q_q is q-th entry of the first structure vector, and n_q is q-th entry of the second structure vector. The third structure vector quantifies the degree of connectedness on each q-level, or in other words, it quantifies the number q-connected components per number of simplices. The values of entries are between 0 and 1, which makes it convenient to compare the density of simplices per connected component on various q-levels. And for our working example, the values of entries are

$$\bar{Q}_q = [0 \quad 0 \quad 0.14 \quad 0.75 \quad 0.88]$$

Simplicial structural complexity

It is obvious so far that simplicial complex may be used as models of complex systems, and hence it would be convenient to quantify complexity based on simplicial complex properties. In the Introduction we already mentioned that we have no shortage with the number of complexity measures, nevertheless for purpose of completeness of simplicial applicability, and necessity to quantify simplicial complexity, we will introduce the so called simplicial structural complexity[13] Ψ.

We have shown that first structure vector stores the information of internal structural properties of simplicial complex. Hence, the quantity which depends on the structure vector and is useful for distinguishing and comparison of different simplicial complexes is *simplicial structural complexity* of complex K introduced by J. Casti[13]:

$$\Psi(K) = \frac{2}{(D+1)(D+2)} \sum_{i=0}^{D} (i+1)Q_i$$

where D is dimension of simplicial complex, and Q_i is the i-th entry of Q-vector. For introducing simplicial structural complexity Casti[13] used three axioms: (1) complexity is 1 for a system build by a single simplex; (2) complexity of a simplicial subcomplex is not greater than complexity of the whole simplicial complex; and (3) the complexity of simplicial complex is not greater than the sum of complexities of its subcomplexes. We will not discuss whether this is a good choice of complexity measure or not, since, as we will see in the following it turned useful for practical reasoning.

It is useful sometimes to calculate and compare the complexities of a simplicial complex and its conjugate, which does not have to be the same. In the case of our working example, it turns out that complexities are actually the same, $\Psi = 2.4$.

Eccentricity

The simplex σ in simplicial complex K is defined by its vertices. If all its vertices are part of some other simplex τ, then simplex σ is completely integrated in another simplex. In terms of simplicial complexes, the simplex is face of some other simplex, which can have equal or larger dimension. Hence, it does not have integrity or individuality in complex as a whole. On the other hand, if simplex does not share vertices with any other simplex, we can say that it is not well integrated or it does have individuality in simplicial complex.

We can define $\breve{q}$ (called bottom q) of simplex σ as the largest dimension of faces which σ shares with other simplices, i.e., the largest q-nearness value. This is equivalent to the value of the q-level on which the simplex firstly connects to some other simplex. We can, also define $\hat{q}$ (called top q) which is equal to the dimension of the simplex. Then the *eccentricity* of simplex σ is defined like[14, 15]:

$$ecc(\sigma) = \frac{\hat{q} - \breve{q}}{\hat{q} + 1}. \tag{2.2}$$

From the definition one sees that $\hat{q}+1$ is the total number of vertices that define simplex σ, and $\hat{q} - \breve{q}$ is the minimal number of vertices which make simplex σ different from the other simplex. Hence, $ecc(\sigma)$ quantifies the individuality or distinguish ability of simplex, and displays the degree of integrity of simplex σ in simplicial complex. The simplex which has $ecc = 0$ is completely integrated into the structure, i.e., the simplex is face of another simplex. The simplex which has $ecc = 1$ does not share vertices (faces) with any other simplex. In the Fig. 2.10 instead of simplices names the values of eccentricity of simplices are written, so that the visual comparison between distinguishability between simplices is made easier.

It is important to mention that equation 2.2 is not the only definition of eccentricity.[12, 14] All of them measure the same properties, and have the mentioned weakness, but the range of values is different. Since the range of ecc defined in 2.2 is [0,1], it was chosen because of practical reasons.

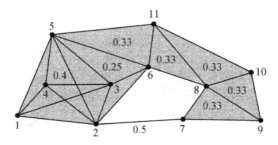

Fig. 2.10 Example of the values of eccentricity for simplices in a complex

Vertex significance

One vertex can be part of many simplices, and we can define a vertex weight θ as the number of simplices to which the vertex belongs. Hence, the vertex weight is the measure that characterizes a vertex. On the other hand, since the simplex is defined by vertices, we can introduce a measure that characterizes a simplex with respect to the vertices that create it. Summing weights of all vertices which create simplex $\sigma_q(i)$, we obtain $\Delta(\sigma_q(i))$. Now we can define *vertex significance*[12, 15] of the simplex with respect to the vertices which create it like:

$$vs(\sigma_q(i)) = \frac{\Delta(\sigma_q(i))}{\max \Delta(\sigma_q(i))} \tag{2.3}$$

where $\max \Delta(\sigma_q(i))$ is the maximal value of all $\Delta(\sigma_q(i))$. The larger values of *vs* indicate larger importance of the simplex with respect to the vertices which create it. As in the case of eccentricity, for our working example in the Fig. 2.11 instead of simplex names the values of vertex significance of simplices are written, so that the visual comparison of significance of simplices with respect to the vertices which build them is made easier.

Notes

With the definition of vertex significance we have finished the introduction of Q-analysis concepts. Nevertheless, it is important to emphasis that with the above quantities the whole potential of Q-analysis apparatus is not exhausted, and we refer an interested

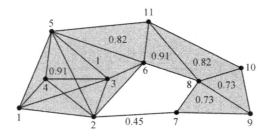

Fig. 2.11   Example of the values of vertex significance for simplices in a complex

reader to the publications of R. Atkin,[5] J. Johnson,[10] and P. Gould *et al.*[14]

2.4.2 *Homology*

Hitherto, structural and connectivity properties of simplicial complex have been explored only through the connectivity of simplices built from the relationship between two sets. In the following we will concentrate on the topological properties of simplicial complex and take into account a key property of the simplicial complex definition: that the power set of the set over which simplicial complex is defined is closed under the formation of subsets. In other words, every sub-simplex, that is the face, is also a simplex in simplicial complex. For the so far considerations it was implicitly accepted, whereas for the introduction of forthcoming concepts it is of an essential importance. Hence, when we say "q-simplices", we mean "all maximal q-dimensional simplices and all q-dimensional faces".

Homology group

We suppose (and hope) that the reader expects that the introduction of new concepts and tools is followed by a convenient illustrative example, hence we will not deprive a reader with it. To show the richness of algebraic topology apparatus, we will use the the same example as before, that is Fig. 2.12.

Let us start again with a finite vertex set $V = \{v_1, v_2, \ldots, v_m\}$. An arbitrary ordering of vertices $\{v_{\alpha_0}, v_{\alpha_1}, \ldots, v_{\alpha_q}\}$ of a simplex

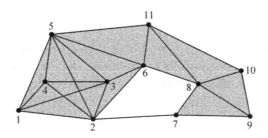

Fig. 2.12   An example of 4-dimensional simplicial complex

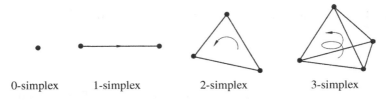

0-simplex   1-simplex   2-simplex   3-simplex

Fig. 2.13   Examples of orientation of 0-, 1-, 2-, and 3-simplex

defines an oriented q-simplex which we denote $[v_{\alpha_0}, v_{\alpha_1}, \ldots, v_{\alpha_q}]$, and we say that simplicial complex K is oriented if all simplices in K are oriented. Note that an unoriented simplex was denoted as $\langle v_{\alpha_0}, v_{\alpha_1}, \ldots, v_{\alpha_q} \rangle$. An example of oriented 0-, 1-, 2-, and 3-simplices is illustrated in Fig. 2.13, and by convention 0-simplex does not have an orientation.

Let $C_q(K)$ (for each $q \geq 0$) be the vector space whose bases is the set of all q-simplices of an oriented simplicial complex K, and the elements are the linear combinations of bases vectors, called *chains*. That is, a *q-chain* is the formal sum of oriented q-simplices

$$c_q = \sum_i a_i \sigma_q(i)$$

where coefficients a_i are the elements of coefficient group, usually the group of integers. Hence, 2-chain is a formal linear combination of triangles, 1-chain is a formal linear combination of edges, and so on $C_q(K)$ is called a *chain group*[1] (the term chain group is accepted for traditional reasons,regardless of the vector space properties of $C_q(K)$, nevertheless $C_q(K)$ is still a group). The dimension of $C_q(K)$ is equal to the q-th entry of an important topological invariant, the *f-vector*, $\boldsymbol{f} = [f_0, f_1, \ldots, f_q, \ldots, f_D]$. In this expression f_q is equal to the number of q-dimensional simplices of the simplicial complex K, i.e., f_0 represents the number of vertices, f_1 number of edges and so on. For q larger than the dimension of K, vector space $C_q(K)$ is trivial and equals to 0. The values of f-vector entries for the example from Fig. 2.12 are

$$\boldsymbol{f} = [11 \quad 24 \quad 18 \quad 6 \quad 1]$$

Note that unlike in the case of vector-like measures introduced in Q-analysis, the first entries of vector-like quantities in homology theory are associated to the 0-dimensional simplices, second entry is associated to 1-dimensional simplices, and so on. Hence, for our example, the simplicial complex contains 11 vertices, 24 edges, 18 triangles, 6 tetrahedra, and 1 pentahedron.

For a set of vector spaces $C_q(K)$ with $0 \le q \le \dim(K)$ the linear transformation $\partial_q : C_q(K) \to C_{q-1}(K)$ called *the boundary operator* acts on the bases vectors $[v_{\alpha_0}, v_{\alpha_1}, \ldots, v_{\alpha_q}]$ in the following way[1]

$$\partial_q [v_{\alpha_0}, v_{\alpha_1}, \ldots, v_{\alpha_q}] = \sum_{i=0}^{q} (-1)^i [v_{\alpha_0}, \ldots, v_{\alpha_{i-1}}, v_{\alpha_{i+1}}, \ldots, v_{\alpha_q}].$$

An example of the action of the boundary operator on a 3-simplex and its subsimplices from Fig. 2.12 is illustrated in Fig. 2.14.

The boundary operator acts on a q-chains by extending its action on q-simplices which form a q-chain, that is

$$\partial_q \left(\sum_i a_i \sigma_i^q \right) = \sum_i a_i \partial_q (\sigma_i^q).$$

Taking a sequence of chain groups $C_q(K)$ connected through the boundary operators ∂_q the so-called *chain complex* is defined in the following way

$$\emptyset \to C_q \xrightarrow{\partial_q} C_{q-1} \xrightarrow{\partial_{q-1}} \cdots \to C_1 \xrightarrow{\partial_1} C_0 \xrightarrow{\partial_0} \emptyset$$

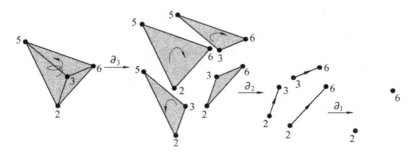

Fig. 2.14   The action of boundary operator on the 3-simplex [2, 3, 5, 6], 2-simplex [2, 3, 6], and 1-simplex [2, 6]

with $\partial_q \partial_{q-1} = \emptyset$ for all q. To illustrate this essential property, take for example a tetrahedron [2, 3, 5, 6] from Fig. 2.14:

$$\partial_3 = [2,3,5,6] = -[2,3,5] + [2,3,6] - [2,5,6] + [3,5,6]$$

and apply boundary operator once more

$$\partial_2\partial_3[2,3,5,6] = -[2,3] + [2,5] - [3,5] + [2,3] - [2,6] + [3,6] - [2,5]$$
$$+ [2,6] - [5,6] + [3,5] - [3,6] + [5,6] = 0.$$

The kernel of ∂_q is the set of q-chains with empty boundary, and a q-cycle (an element of the *group of cycles* Z_q) is a q-chain in the kernel of ∂_q. In other words, the boundary operator maps q-chains from Z_q to zero. The image of ∂_{q+1} is the set of q-chains which are boundaries of $(q + 1)$-chains, denoted by B_q (called the *group of boundaries*). The groups of cycles Z_q and of boundaries B_q are the subgroups of chain group C_q, that is $B_q \subseteq Z_q \subseteq C_q$. The *q-th homology group*[1] is defined as

$$H_q = \frac{\ker \partial_q}{\mathrm{im}\partial_{q+1}} = \frac{Z_q}{B_q}.$$

The elements of homology group H_q are equivalence classes of q-cycles which are not boundaries of any $(q + 1)$-chain, and can be understood intuitively that homology characterizes q-dimensional holes. The rank of the q-th homology group $\beta_q = \mathrm{rank}(H_q)$ or $\beta_q = \dim(H_q)$ is topological invariant called the q-th *Betti number* and is equal to the number of q-dimensional holes in simplicial complex. Since it is a topological invariant it is used to distinguish topological spaces one from another. For example, the value of β_0 is the number of connected components of simplicial complex, β_1 is the number of tunnels, β_2 is the number of voids, etc. In the simplicial complex presented in Fig. 2.12 there is one connected component, hence $\beta_0 = 1$, and one 1-dimensional hole bounded by 1-dimensional simplices [2,6], [2,7], [7,8], and [6,8], hence $\beta_1 = 1$, as illustrated in Fig. 2.15.

Betti numbers, and f-vector are related to another important topological invariant, called the *Euler characteristic*.[2] Namely, for

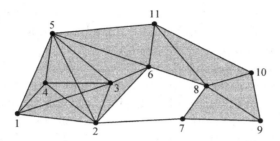

Fig. 2.15   Examples of two 1-dimensional chains in 4-dimensional simplicial complex

a simplicial complex K, with f-vector values

$$\boldsymbol{f} = [f_0, f_1, \ldots, f_q, \ldots, f_D]$$

the Euler characteristic is defined as

$$\chi = \sum_{i=0}^{D} (-1)^i f_i$$

and from the *Euler-Poincaré theorem*, alternating sum of Betti numbers is the same as the Euler characteristic, that is

$$\chi = \sum_{i=0}^{D} (-1)^i \beta_i.$$

As R. Atkin pointed out in[5] the zeroth Betti number is equal to Q_0, nevertheless, the higher-order Betti numbers are not equal to the higher order Q-vector entries. Therefore, the analysis presented in the previous section (Q-vector) gives a generalization of the zeroth order Betti number, although different from the homology theory. On the other hand, since q-levels represent division of a simplicial complex into subcomplexes K_q, we can calculate homology groups of each subcomplex and obtain an additional insight into properties of simplicial substructure. To our knowledge such calculations have not been performed so far, and we suggest a reader to perform calculation as a practice. As Dowker[3] have proved, the homology groups of simplicial complex and its conjugate complex are isomorphic, therefore the values of Betti numbers of simplicial complex are preserved in its conjugate complex.

Each boundary operator ∂_q has its matrix representation B_q with respect to bases of vector spaces $C_q(K)$ and $C_{q-1}(K)$, with rows associated with the number of $(q\text{-}1)$-simplices and the columns associated with the number of q-simplices. To each boundary operator ∂_q corresponds an adjoint operator $\partial_q^* : C_{q-1}(K) \to C_q(K)$ with the associated matrix rep resentation equal to the transpose of matrix representation of boundary operator ∂_q, that is B_q^T. It is important to mention that the q-th adjoint boundary operator is in fact the same as the q-th coboundary operator $\delta_q : C^{q-1}(K) \to C^q(K)$,[1] whereas, their matrix representations coincide when proper scalar products are chosen for the definitionof ∂_q^*. The usefulness of as well as the relation between matrix representations of boundary and coboundary operators will be more clear when they are used in the section related to combinatorial Laplacian.

Persistent homology

So far we have introduced homology groups and topological invariants related to it, and now we can extend the concept of homology to the methods which capture the homological changes of a space that is undergoing growth. The intuition that lies behind introducing persistent homology[16] is rather simple: recorded changes that topological space suffer encoded in the change of homology groups may help us to infer something about that topological space. In other words, we want to track the changes of homology groups during the changes of simplicial complex, and therefore get some additional information about the simplicial complex. The changes of simplicial complex are usually followed by changes of some free parameter, and since in this section the persistent homology will be introduced in general, for now we will not specify the meaning of free parameter. Although the beginning of section sounds too blurry, to diminish blurriness we will, as usual, illustrate the ideas with simple examples. In Fig. 2.16(a) we can see how the structure of a simplicial complex is changing

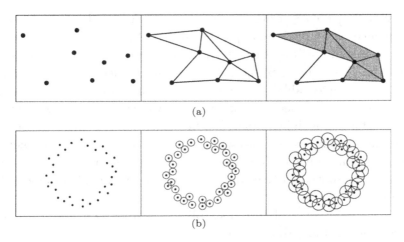

Fig. 2.16 Example of topological changes of simplicial complex (a), and set of points (b)

by adding new vertices, edges, and triangles. On the other hand, in Fig. 2.16(b) we can see a set of points forming approximately the shape of some circular object, say annulus. But in order to reconstruct the topological shape we need to do an additional task: take each of points as centers of equal-sized discs and let the radius of discs grow. From a particular value of radius the union of overlapping discs will recover the shape (in topological sense) of a circular object.

In both examples we can notice that some objects (like connected components or holes) appear and disappear, whereas some are long-lived no matter how much we change a parameter (time in Fig. 2.16(a) example, and radius in Fig. 2.16(b) example). Of course, increasing the radius in the Fig. 2.16(b) example the hole in center will be filled in, but the point is that the hole in the center was preserved the longest.

After these intuitive examples it is easy now to generalize the concept of persistent homology. Hence, in the broadest sense persistent homology records topological properties of a nested sequence of spaces (in our case simplicial complexes) called *filtration*[17]

$$\emptyset = K_{r_0} \subseteq K_{r_1} \subseteq \cdots \subseteq K_{r_n} = K \qquad (2.4)$$

where $r_i \leq r_j$ for $i \leq j$ are free index parameters. The nestedness of a sequence of simplicial complexes implies that we are dealing with a sequence of subcomplexes which are constructred in the process of filtration. From the examples, it is easy to deduce that free parameter r_i may take values of time or radius or diameter or anything, depending on the specific system which we want to analyze. Or filtration of a simplicial complex can be understood as the evolution of a simplicial complex or a growing sequence of simplicial complexes, implying that during filtration stages simplices are always added, but never removed, hence, further implying a partial order on the simplices.

Hence, a sequence of nested simplicial complex (2.4) induces maps on homology of any dimension k,

$$0 \to H_k(K_{r_1}) \to H_k(K_{r_2}) \to \cdots \to H_k(K_{r_n}).$$

Since, by definition, the homology enumerate the equivalent classes of cycles by factoring out the boundary cycles, hence, the focus is on the enumeration of nonbounding cycles whose life-span lasts beyond parameter threshold, and which determine persistent or long-lasting topological properties of the complex. Hence, the important cycles persist through the long sequence of parameter r_i values. The persistent homology group $H_k(r_i, r_j)$ quantifies the topological invariants from K_{r_i} that are still present in K_{r_j}. In other words, it is defined by taking into consideration cycles in K_{r_i} to be equivalent with respect to the boundaries in K_{r_j}, i.e.,

$$H_k(r_i, r_j) = \frac{Z_k(r_i)}{B_k(r_j) \cap Z_k(r_i)}$$

Hence, persistent homology deliver us an integrative picture of the topological structure present at all moments of filtration, and thus tracks the *birth* (*appearance*) and *death* (*disappearance*) of every equivalence class of cycles, that is, homology group generators. A homology group generator in $H_k(K_{r_i})$ is said to persist if its image in $H_k(K_{r_{i+1}})$ is also nonzero, or otherwise it is said that it is dead, and homology group generator in $H_k(K_{r_j})$ is said to be born when it is not in the image of $H_k(K_{r_{j-1}})$. Let us elaborate this rather vague statement. By changing (i.e., increasing) a free parameter, like in the above examples, new simplices are added into the simplicial complex. On the other hand, recall that the generators of homology groups are, in simple words, the holes, which may be understood as parts of simplicial complex where simplices are missing. As we go from one filtration stage to another adding new simplices may destroy some homology generators, i.e., may "fill" holes, or it may create new homology generator, i.e., may create a new hole. The filtration stage at which a particular homology group generator is created is called the *birth stage*, labeled by r_b, whereas the filtration stage at which it is destroyed is called the *death stage*, labeled by r_d. Hence, the births and deaths correspond to changes in the topology of complex through filtration. The outcome of computing filtration stages in persistent homology is the list of pairs (r_b, r_d) that represent the birth and death of each homology class in the filtration, in other words, the pair of indices (r_b, r_d) for each homology generator represent persistence interval of each homology group generator.

Although there are different ways to represent graphically the results of computing persistent homology, we will restrict ourselves on two most common: the barcode[18, 19] and the persistence diagram.[20] In persistence barcode the horizontal axes represents the change of filtration parameter, vertical axes represents homology group generators, and the persistence interval (or bar) is represented as a horizontal line associated to the p-th homology generator's birth/death filtration stage, spanned from r_b to r_d. Hence, for each p we have a sequence of horizontal lines with different lengths (Fig. 2.17). The persistence barcode encodes the description of topological evolution: the length of an interval indicates the significance of a hole, and hence the shorter interval, the less significant is a hole; long bars indicate robust topological features with respect to the changes of a parameter. In the persistence diagram abscissa is associated with the free parameter values of the birth of homology generators, and ordinate is associated with the free parameter values of the death of homology generators. Hence, the coordinates of a point in persistence diagram (r_b, r_d) represent birth and death of a particular generator (i.e., a q-dimensional hole) of the q-th homology group. Notice that

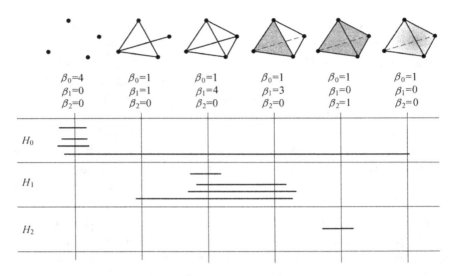

Fig. 2.17   An example of filtration of simplicial complex represented as a single simplex — tetrahedron, together with corresponding barcode

the points in persistence diagram lie above the diagonal. Short bars in the persistence barcode and the points on and near diagonal in persistence diagram are associated with short-lived q-dimensional holes and they represent the *topological noise*.[21]

In the Fig. 2.17 we illustrated an arbitrary filtration stages of a simplicial complex built by only one simplex-tetrahedron. For simplicity we have assumed that the criteria for building a simplicial complex may be neglected, and we will focus on the topological changes of a simplex through the filtration process. Hence, if at the first filtration stage we add only four vertices, i.e., 0-simplices (colored in gray in Fig. 2.17) we have four disconnected components ($\beta_0 = 4$). At the next stage the disconnectivity is destroyed by adding four edges, i.e., 1-simplices (colored in gray in Fig. 2.17), and hence the zeroth Betti number decreases, resulting in $\beta_0 = 1$. Nevertheless, at this stage one non-boundary cycle, that is a 1-dimensional hole, emerges, increasing first Betti number: $\beta_1 = 1$. Even at these early stages we can notice that addition of new simplices destroy some invariants and build new ones. But these two precesses does not have to occur at the same time, since at the next stage we have only the birth of holes by adding two more edges, i.e., 1-simplices (colored in gray in Fig. 2.17), and increase of first Betti number ($\beta_1 = 1 \rightarrow \beta_1 = 4$). The next stage is followed by addition of a triangle, i.e., 2-simplex (colored in gray in Fig. 2.17), which unlike in the previous stage, destroys a hole, and hence first Betti number decreases ($\beta_1 = 4 \rightarrow \beta_1 = 3$). In the fifth stage more triangles (2-simplices colored in gray in Fig. 2.17) are added, influencing the destruction of 1-dimensional holes ($\beta_1 = 3 \rightarrow \beta_1 = 0$), and building a new 2-dimensional hole, hence increasing second Betti number ($\beta_2 = 0 \rightarrow \beta_2 = 1$). This stage can be understood like building an empty generalized pyramid or an empty tetrahedron. Finally, at the last stage the 3-simplex is added (colored in gray in Fig. 2.17) giving birth of a filled tetrahedron, and destroying 2-dimensional hole ($\beta_2 = 1 \rightarrow \beta_2 = 0$). Alongside with this analysis of transitions between filtration stages, we can notice the emergence the short bars, that is short-lived holes, which does not have a significant influence on the final structure of a simplicial complex. Even from this simple example we can get a hint of usefulness of persistent homology, and when we proceed with the application on the non-trivial simplicial complex it will be even more obvious.

A barcode for a more complex simplicial structure illustrated in Fig. 2.18 exemplifies a snapshots of filtration stages, neglecting the criteria for addition of simplices. Like in the previous example, the newly added simplices are colored in red. Although the figure illustrates only a sample of filtration stages, from visual inspection it is evident that the amount of topological noise can be significant, and that especially at the early filtration stages may mask persistent topological invariants. For example, many short bars at the first stage completely mask the existence of a persistent 1-dimensional hole, obviously present at end of filtration process.

Even from a simple visual inspection of the examples, we can deduce that the rank of persistent homology group $H_k^{r_i \rightarrow r_j}$ equals the number of intervals in the barcode of homology group H_k within the limits of the corresponding parameter range, i.e., lifetime $[r_i, r_j]$, where r_i and r_j may be associated to the filtration complexes $K_i(r_i)$ and $K_j(r_j)$, respectively.

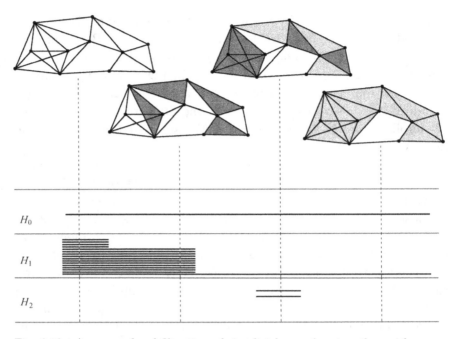

Fig. 2.18   An example of filtration of simplicial complex, together with corresponding barcode

Combinatorial Laplacian

After reading a section about persistent homology, the reader may be surprised that we proceed with the introduction of another algebraic topology concept, which maybe at the first sight does not involve homology. Nevertheless, at the second sight it is tightly related to homology, since it may be used for calculation of a rank of homology groups, that is, Betti numbers, and hence it may be used as a helpful computational tool. The content of this section relies on the T. Goldberg's thorough introductory manuscript "Combinatorial Laplacians of Simplicial Complexes".[22]

From a definition of simplicial complex, any face of a simplex is a simplex as well. Hence, let us take two q-simplices, say σ_q and τ_q. They can both be faces of some higher dimensional simplex, say $(q+1)$-simplex, implying that these two simplices are neighbors through membership to the same higher-dimensional simplex. On the other hand, these two simplices can share a lower-dimensional simplex (that is share a face) say $(q-1)$-simplex, implying now that they are neighbors through sharing a common face. This intuitive illustration of non-trivial adjacency between simplices gives us a new task: how to store and use this information? For derivation of an appropriate quantity we will rely on the so far introduced concepts of oriented simplex and boundary operator. Recall that an arbitrary ordering of vertices of a q-simplex defines an oriented q-simplex, whereas a boundary operator maps q-simplices on $(q-1)$-simplices.

For defining an appropriate quantity the intuition is not enough, so we will give a more formal definition, and at the same time easy to follow exposition. Two q-simplices $\sigma_q(i)$ and $\sigma_q(j)$ of an oriented simplicial complex i are *upper adjacent*, denoted $\sigma_q(i) \sim_U \sigma_q(j)$, if they are both faces of some $(q+1)$-simplex in K. The *upper degree* of a q-simplex σ_q in K, denoted $\deg_U(\sigma_q)$, is the number of $(q+1)$-simplices in K of which σ is a face. If oriented q-simplices $\sigma_q(i)$ and $\sigma_q(j)$ are upper adjacent and have a common $(q+1)$-simplex τ, we say that $\sigma_q(i)$ and $\sigma_q(j)$ are *similarly oriented* if orientations of $\sigma_q(i)$ and $\sigma_q(j)$ agree with the ones induced by τ. For two q-simplices $\sigma_q(i)$ and $\sigma_q(j)$ of an oriented simplicial complex K we say that they are *lower adjacent*, denoted $\sigma_q(i) \sim_L \sigma_q(j)$, if they have common $(q-1)$-face (that is $(q-1)$-simplex as a face). Hence, the *lower degree* $(\deg_L(\sigma_q))$ of a q-simplex is defined as the number of $(q-1)$-faces in σ_q, which is always equal to $q+1$.

For easier understanding of upper/lower adjacency, let's pay attention on the Fig. 2.19. Two tetrahedra, i.e., 3-simplices, $\langle 0,1,2,3 \rangle$ and $\langle 1,2,3,4 \rangle$ share a triangle, i.e., 2-dimensional, simplex (2-face) $\langle 1,2,3 \rangle$, and therefore we say that they are lower adjacent. On the other hand, two triangles, i.e., 2-simplices, $\langle 0,1,2 \rangle$ and $\langle 1,2,3 \rangle$ are both faces of a tetrahedron, i.e., 3-simplex, $\langle 0,1,2,3 \rangle$, and hence they are upper adjacent. The upper degree of a triangle $\langle 1,2,3 \rangle$ is $\deg_U(\langle 1,2,3 \rangle) = 2$ since it is a face of two tetrahedra $\langle 0,1,2,3 \rangle$ and $\langle 1,2,3,4 \rangle$, whereas the lower degrees of a tetrahedral $\langle 0,1,2,3 \rangle$ and $\langle 1,2,3,4 \rangle$ is the same and $\deg_L(\langle 0,1,2,3 \rangle) = \deg_L(\langle 1,2,3,4 \rangle) = 4$, since each of them contains four triangles.

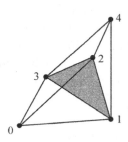

3-simplices <0,1,2,3> and <1,2,3,4> are lower adjacent since they share common 2-fcae <1,2,3>

2-simplices <0,1,2> and <1,2,3> are upper adjacent since they are both faces of 3-simplex <0,1,2,3>

Fig. 2.19   An example of adjacency between two 3-simplices

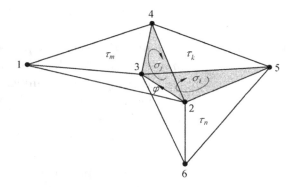

Fig. 2.20   An example of upper and lower adjacency between simplices $\sigma_q(i)$ and $\sigma_q(j)$

In the little more complicated example in Fig. 2.20 the idea of upper/lower adjacency is even more clear. Simplices $\sigma_q(i)$ and $\sigma_q(j)$ have twofold adjacency: they are upper adjacent through membership to the simplex τ_k, and lower adjacent through sharing the simplex φ.

Defining the boundary operator and its adjoint we have provided necessary conditions for the definition of combinatorial Laplacian of simplicial complex. Namely, for a simplicial complex K and an integer $q \geq 0$, the q-th *combinatorial Laplacian* is linear operator (since the composition of linear maps is a linear map) defined as $L_q : C_q \rightarrow C_q$ and given by[23]

$$L_q = \partial_{q+1} \circ \partial_{q+1}^* + \partial_q^* \circ \partial_q \tag{2.5}$$

Relating upper/lower adjacency and the definition (2.5) we come to the convenient notation:

$$L_q^{UP} = \partial_{q+1} \circ \partial_{q+1}^* \quad \text{and} \quad L_q^{DN} = \partial_q^* \circ \partial_q$$

where L_q^{UP} is referred to as the upper combinatorial Laplacian and L_q^{DN} is the down combinatorial Laplacian. Corresponding matrix representation relative to some ordering of the standard bases for C_q and C_{q-1} for the q-th *Laplacian matrix* of K is

$$L_q = B_{q+1}B_{q+1}^T + B_q^T B_q.$$

As in the case of the Laplacian operator we may use the following notation for convenience:

$$L_q^{UP} = B_{q+1}B_{q+1}^T \quad \text{and} \quad L_q^{DN} = B_q^T B_q.$$

As an illustrative example, let us calculate combinatorial Laplacians of simplicial complex in Fig. 2.21. The matrix representation of boundary operators has the following form:

$$\boldsymbol{B}_1 = \begin{bmatrix} -1 & 0 & -1 & 0 \\ 1 & -1 & 0 & -1 \\ 0 & 1 & 1 & 0 \\ 0 & 0 & 0 & 1 \end{bmatrix}$$

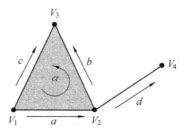

Fig. 2.21   An oriented simplicial complex in which every simplex is labeled

where rows are associated to $v_1 \sim v_4$, and columns are associated to $a \sim d$, and

$$B_2 = \begin{bmatrix} 1 \\ 1 \\ -1 \\ 0 \end{bmatrix}$$

where rows are associated to $a \sim d$, and column is associated to α.

It is easy to read these matrices: for example, in the matrix B_1 an element $B_1(1,1)$ is equal to -1 since the vertex v_1 is a source in 1-simplex α, and an element $B_2(2,1)$ is equal to 1 since the vertex v_2 is a sink in 1-simplex α, or in the matrix B_2 an element $B_2(1,1)$ is equal to 1 since an orientation of triangle α is the same as the orientation of an edge α, where as an element $B_2(3,1)$ is equal to -1 since an orientation of triangle α has the opposite orientation of an edge c. The q-th combinatorial Laplacians for this simplicial complex are:

$$L_0 = B_1 \cdot B_1^T, L_1 = B_2 B_2^T + B_1^T B_1$$
$$L_2 = B_2^T B_2$$

We leave to the reader as a simple exercise the explicit calculation of the above combinatorial Laplacians, as well as to separate matrices of upper and lower combinatorial Laplacians.

We are now equipped with enough material to proceed further, and reveal the meaning of combinatorial Laplacian, which emerge from the capturing of upper/lower adjacency.[24] Clearly, graph is a 1-dimensional simplicial complex since links (1-dim simplices) connect nodes (0-dimensional simplices) and the largest dimension of a simplex in the complex is 1. Hence, we can start with the derivation of a combinatorial Laplacian. The 0-dimensional combinatorial Laplacian of simplicial complex K is a linear map $L_0 : C_0(K) \rightarrow C_0(K)$, and since the maps ∂_0 and ∂_0^* are assumed to be zero maps, it follows that

$$L_0 = \partial_1 \circ \partial_1^*$$

where the boundary operator $\partial_1 : C_1(K) \rightarrow C_0(K)$ maps edges to vertices. Since in the matrix representation B_1 of boundary operator

∂_1 the rows are associated with vertices and the columns are associated with edges (see previous example related to the Fig. 2.21), it is obvious that the matrix $\boldsymbol{B}_1$ is equal to the incidence matrix of an oriented graph.[25] Therefore matrix representation of combinatorial Laplacian is $\boldsymbol{L}_0 = \boldsymbol{B}_1 \cdot \boldsymbol{B}_1^T$, and the matrix elements are

$$(L_0)_{ij} = \begin{cases} \deg(v_i) & \text{if} \quad i = j \\ -1 & \text{if} \quad v_i \sim v_j \\ 0 & \text{otherwise} \end{cases} \tag{2.6}$$

where $\deg(v_i)$ is vertex degree (that is number of neighbors of a vertex v_i) and the relation $v_i \sim v_j$ is the adjacency relation between vertices v_i and v_j, and is the same as upper adjacency $v_i \sim_U v_j$. Clearly, the entries of the 0-dimensional combinatorial Laplacian are the same as the graph Laplacian entries defined in the usual way via expression $\boldsymbol{L}_{graph} = \boldsymbol{D} - \boldsymbol{A}$, where diagonal entries of matrix $\boldsymbol{D}$ are equal to the vertex degrees ($D_{ii} = \deg(v_i)$) and nondiagonal entries are zeros, and the entries of matrix A are $A_{ij} = 1$ if $v_i \sim v_j$, $A_{ij} = 1$ if vertices v_i and v_j are not adjacent, and $A_{ii} = 0$ (undirected, unweighted, without loops and multiple edges graph).[26]

For the general case let us assume that K is an oriented simplicial complex, q is an integer with $0 < q \leq \dim(K)$, and let $\{\sigma^1, \sigma^2, \dots, \sigma^n\}$ denote the q-simplices of complex K, then it is not difficult to deduce from $L_q = L_q^{UP} + L_q^{DN}$ that

$$(L_q)_{ij} = \begin{cases} \deg_U(\sigma^i) + q + 1 & \text{if } i = j \\ 1 & \text{if } i \neq j \text{ and } \sigma^i \text{ and } \sigma^j \text{ are not upper} \\ & \text{adjacent but have a similar common} \\ & \text{lower simplex} \\ -1 & \text{if } i \neq j \text{ and } \sigma^i \text{ and } \sigma^j \text{ are not upper} \\ & \text{adjacent but have a dissimilar common} \\ & \text{lower simplex} \\ 0 & \text{if } i \neq j \text{ and } \sigma^i \text{ and } \sigma^j \text{ are upper} \\ & \text{adjacent or are not lower adjacent} \end{cases} \tag{2.7}$$

since $(L_q^{UP})_{ii} = \deg_U(\sigma^i)$ and $(L_q^{DN})_{ii} = \deg_L(\sigma^i)$.

Detailed proof of the above expression is straightforward.[22] For later use it would be useful to notice that $(L_q)_{ii} = \deg_U(\sigma^i) + \deg_L(\sigma^i) = \deg_U(\sigma^i) + q + 1$ since every simplex of dimension $q > 0$ has exactly $q + 1$ $(q-1)$-faces. Clearly, for $q = 0$ Laplacian matrix of general simplicial complex reduces to graph Laplacian.

The above definition of matrix elements is unhandy and impractical for applications of large simplicial complex, and hence we need to develop some computationally convenient way to extract useful and meaningful information from combinatorial Laplacians. In that course, let us focus now on the eigenvalues and eigenvectors of q-th combinatorial Laplacian L_q. For an oriented simplicial complex K and an integer q with $0 \leq q \leq \dim(K)$, the q-th *Laplacian spectrum* is denoted as $S(L_q(K))$. It represents set of eigenvalues of $L_q(K)$ together with their multiplicities and is independent on the choice of orientation of q-simplices in the complex K. Since the q-th Laplacian matrix is positive semidefinite, all its eigenvalues are nonnegative. The null space of $N(L_q(K))$ is the eigenspace of $L_q(K)$ and corresponds to the zero eigenvalues. At this moment, as we promised, we can relate the combinatorial Laplacian and homology: the combinatorial Hodge theorem states that the q-th homology group $H_q(K)$ is isomorphic to the null space of q-th combinatorial Laplacian,[27] that is

$$H_q(K) \cong N(L_q(K))$$

for each integer q with $0 \leq q \leq \dim(K)$. Therefore, the multiplicity of zero eigenvalues of q-th combinatorial Laplacianis equal to the number of the q-dimensional holes in a simplicial complex, i.e., a Betti number. This is very useful expression providing a practical method for calculation of Betti numbers.[28]

In the following we will introduce some properties of the spectra of the q-th combinatorial Laplacian which will be useful for the analysis and interpretation of results. If simplicial complex K consists of disconnected components which are themselves simplicial complexes $K_1, K_2, \ldots, K_n$, then the spectra of q-th combinatorial Laplacian $L_q(K)$ of K for each q with $0 \leq q \leq \dim(K)$ are equal to the union of spectra of each $L_q(K_i)$ for $i = 1, \ldots, n$ separately,[22] that is

$$S(L_q(K)) = S(L_q(K_1)) \cup S(L_q(K_2)) \cup \cdots \cup S(L_q(K_n))$$

Another very important property is that if simplicial complex K is formed by gluing two simplicial complexes K_1 and K_2 along a q-face, then the spectrum S is the union of spectra of K_1 and K_2, i.e., $S(L_i(K)) = S(L_i(K_1)) \cup S(L_i(K_2))$ for all $i \geq q + 2$.[22] Since we will consider examples of simplicial complexes formed by the cliques of a graph (i.e., collections of nodes all connected with each other), we want to emphasis that the spectrum of a single k-clique (a clique containing k vertices), denoted by G, is $S(L_0(G)) = \{0, [k]^{k-1}\}$,[29] which is equivalent to $S(L_0(G)) = \{0, [k]^{k_0-1}\}$, and $S(L_i(G)) = \{0, [k]^{f_i-1}\}$, where $i = 2, \ldots, k$, and $f_0, f_1, \ldots, f_{k-1}$ are the entries of f-vector, and the exponent of $[k]$ means the multiplicity of an eigenvalue k. These properties are consequences of (2.6) and (2.7). Namely, every vertex in a k-clique G has upper degree $k-1$ and every pair of distinct vertices has a dissimilar common lower simplex (an edge), hence for $q = 0$ from (2.6) implies that

$$\mathbf{L_0}(G) = \begin{pmatrix} k-1 & -1 & \cdots & -1 \\ -1 & k-1 & \cdots & -1 \\ \vdots & \vdots & & \vdots \\ -1 & -1 & \cdots & k-1 \end{pmatrix}.$$

Hence solving the eigenvalue problem of $\mathbf{L}_q(G)$ implies that $S(L_0(G)) = \{0, [k]^{k-1}\}$, and since the 0-th entry of f-vector is equal to the number of vertices in a complex $f_0 = k$, we can write the general expression $S(L_0(G)) = \{0, [k]^{f_0-1}\}$. For $(k-1) \geq q > 0$, every q-simplex σ^i in G has upper degree equal to $\deg_U(\sigma^i) = (k-1) - q$ and every pair of distinct q-simplices σ^i and σ^j are upper adjacent, hence from (2.7) implies that for $q > 0$

$$\mathbf{L}_q(G) = \begin{pmatrix} k & 0 & \cdots & 0 \\ 0 & k & \cdots & 0 \\ \vdots & \vdots & & \vdots \\ 0 & 0 & \cdots & k \end{pmatrix}_{f_q \times f_q}.$$

The eigenvalue spectra has only one eigenvalue $\lambda = k$ with multiplicity equal to the number of q-simplices which is equal to f_q, the q-th entry of f-vector, so that $S(L_q(G)) = \{[k]^{f_q}\}$. For a single $(k-1)$-simplex f_q is equal to the number of q-dimensional faces, that is

$$f_q = \frac{k!}{(k-1-q)!(q+1)!}.$$

An example of the properties of combinatorial Laplacian spectra is illustrated in Fig. 2.22 where Sq denotes the q-th component of the spectrum. In (a) the complex consisting of two disjointed simplices is presented with the corresponding spectra and in (b) through (d) the two simplices are first joined along a 0-dimensional face (case(b)), followed by attachment along a 1-dimensional face (case(c)) and completing the process with attachment along a 2-dimensional face.

In order to utilize more practical methods for comparison of the eigenvalue spectra of combinatorial Laplacian for different simplicial complexes, most transparent mode is visualization. To avoid problems which emerge from histogram or relative frequency plots due to the choice of the number of bins and their size and since we are dealing with $\dim(K)+1$ eigenvalue spectra for a single simplicial complex (a fairly large number), we need a visualization method which depends on a single valued parameter unique for all the plots. For that purpose can be used the convolution of the spectral density represented by Dirac delta function $\sum_i \delta(\lambda, \lambda_q^i)$ with a smooth kernel $g(x, \lambda)$ so that the density function[30]

$$f(x) = \int g(x, \lambda) \sum_i \delta(\lambda, \lambda_q^i) d\lambda = \sum_i g(x, \lambda_q^i)$$

has advantageous visual properties. In the above expression λ_q^i is i-th eigenvalue of the q-th combinatorial Laplacian. Many kernels may be rendered useful in forming the density function, such as the Cauchy-Lorentz distribution $\frac{1}{\pi} \frac{\gamma}{(\lambda-x)^2+\gamma^2}$ or the Gaussian distribution $\frac{1}{\sqrt{2\pi}\sigma} \exp\left(-\frac{(x-m_x)^2}{2\sigma^2}\right)$. For considering the examples in current book, our choice is the Cauchy-Lorentz kernel yielding the

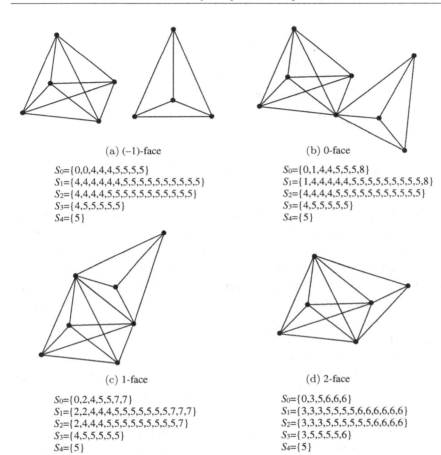

(a) (−1)-face

S_0={0,0,4,4,4,5,5,5,5}
S_1={4,4,4,4,4,4,5,5,5,5,5,5,5,5,5}
S_2={4,4,4,4,5,5,5,5,5,5,5,5,5,5,5}
S_3={4,5,5,5,5,5}
S_4={5}

(b) 0-face

S_0={0,1,4,4,5,5,5,8}
S_1={1,4,4,4,4,4,5,5,5,5,5,5,5,5,5,8}
S_2={4,4,4,4,5,5,5,5,5,5,5,5,5,5,5}
S_3={4,5,5,5,5,5}
S_4={5}

(c) 1-face

S_0={0,2,4,5,5,7,7}
S_1={2,2,4,4,4,5,5,5,5,5,5,5,7,7,7}
S_2={2,4,4,4,5,5,5,5,5,5,5,5,5,7}
S_3={4,5,5,5,5,5}
S_4={5}

(d) 2-face

S_0={0,3,5,6,6,6}
S_1={3,3,3,5,5,5,5,6,6,6,6,6,6}
S_2={3,3,3,5,5,5,5,5,5,6,6,6,6}
S_3={3,5,5,5,5,6}
S_4={5}

Fig. 2.22 Spectrum of simplicial complex formed by 4-simplex and 3-simplex, when they share: (a) (−1)-face; (b) 0-face; (c) 1-face; (d) 2-face

following density function

$$f(x) = \sum_i \frac{\gamma}{(\lambda_q^i - x)^2 + \gamma^2}$$

where γ is a fixed parameter which regulates the resolution (the level of detail in the plot) so that a too high value blurs the spectrum while too low value disguises it. In all spectra presented in examples the value $\gamma = 0.03$ was used.

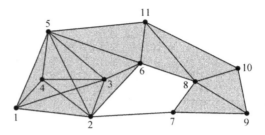

Fig. 2.23   An example of 4-dimensional simplicial complex

For illustration of the above visualization method we will return to our working example of simplicial complex (Fig. 2.23). The convoluted spectra of combinatorial Laplacians are presented in Fig. 2.24 *E0-E4*, where peaks are associated to the eigenvalues. Comparing the simplicial complex in Fig. 2.23 and the eigenvalues for combinatorial Laplacians L_0 and L_1 we can notice the emergence of eigenvalues equal to zero, which originate from the existence of a single connected component, and 1-dimensional hole, respectively. This result confirms the combinatorial Hodge theorem, that the null space of q-th combinatorial Laplacian L_q is isomorphic to the q-th homology group H_q, and hence the multiplicity of eigenvalues equal to zero is the same as Betti number. In some sense, we expected this result, and we only confirmed the combinatorial Hodge theorem in practice, but let us now seek for some additional information which can be extracted from this data. The emergence of a peak for the eigenvalue equal to 5 is noticeable. Namely, this eigenvalue appears for the zeroth combinatorial Laplacian and persists all to the fourth combinatorial Laplacian. From Fig. 2.23 we see that the largest simplex is 4-simplex, formed by 5 vertices, and when we recall the properties of spectra of combinatorial Laplacian presented above, the interpretation of the result is easy: the eigenvalue equal to 5 originates from the existence of a 4-simplex in complex, and its connectivity with other simplices. We leave to the reader as a simple exercise to explicitly relate this result with the properties of combinatorial Laplacian spectra. Alongside with the dominance of this particular eigenvalue, we can notice that, as the dimension increases, the eigenvalues away

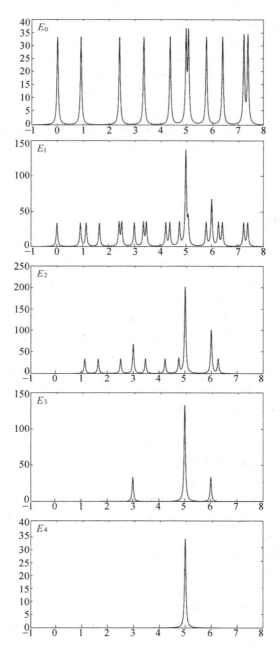

Fig. 2.24   Spectra of combinatorial Laplacians for dimensions 0~4

from the eigenvalue equal to 5 disappear, leaving that eigenvalue alone at 4-th dimension. This result can be interpreted as the indication of a significance, as well as dominance, of a particular simplex. This result maybe looks too trivial, since by a simple inspection of a Fig. 2.23 we can conclude the same. Nevertheless, we are often working with large simplicial complexes which cannot be presented graphically as in our working example, and hence development of methods which can uncover some not so obvious information, are of the paramount importance.

2.5   Summary

This very brief introduction of algebraic topology of simplicial complexes does not cover many other interesting and important topics. Nevertheless, the authors tried to give an introduction which will serve as a starting point for research and additional learning for an interested reader, on the one hand, and to highlight current computational tools which emerge from algebraic topology, on the other hand.

Hence, what have we learnt so far? Simplicial complexes can be defined in different ways (geometrically, abstractly, and from a relation), and still keep the similar properties, meaning that all three definitions are equivalent. Of course, depending on the underlying problem of application, we can choose a particular definition, and then shift from one definition to another. Building a simplicial complex is just a part of the job — we need some tools and methods to analyze it. We gave an introduction to the Q-analysis, and many intuitive and practical quantities which emerge from Q-analysis. Although the research in Q-analysis as a tool gave many results, we believe that that research area has a large potential for further developments. Nevertheless, unlike the Q-analysis, the research in homology is exceptionally fertile now days, as can be seen from this short introduction. We have introduced the basic homological concepts, like boundary operators, chain groups, cyclic groups, boundary groups, homology groups, Betti numbers. And these concepts were not enough, but we introduced a persistent homology and combinatorial Laplacian, both emerging from homological concepts. Through whole chapter we calculated the introduced quantities of the same example. The intention was three fold: to help reader to better understand the concepts, then to show the versatility of analysis applied on the same simplicial complex, and finally, for an interested reader to redo all the calculations by himself/herself as an exercise.

The working example is obviously a made up simplicial complex, constructed with the purpose to highlight all important characteristics of algebraic topology concepts which are introduced quite generally, without a reference on any real-world application. Hence, we need to proceed to the next step, which will connect these abstract definitions, and real-world phenomena. That next step leads us to the next chapter where we will introduce the ways of building simplicial complexes from different data. Even from the so far exposition we can have hints of possible application, nevertheless, the procedure, as well as criteria, for building a simplicial complex from data often is not trivial, and it is influenced by the characteristics of the underlying problem.

Bibliography

[1] MUNKRES J R. Elements of algebraic topology [M]. California: Addison-Wesley Publishing, 1984.
[2] KOZLOV D. Combinatorial algebraic topology [M]. Heidelberg: Algorithms and Computation in Mathematics, Springer-Verlag, 2008.
[3] DOWKER C H. Homology groups of relations [J]. Annals of Mathematics, 1952, 56: 84.
[4] ATKIN R H. From cohomology in physics to q-connectivity in social sciences [J]. Int. J. Man-Machine Studies, 1972, 4: 341.
[5] ATKIN R H. Combinatorial connectivities in social systems [M]. Stuttgart: Birkhäuser Verlag, 1977.
[6] ATKIN R H. Mathematical structure in human affairs [M]. London: Heinemann, 1974.
[7] JOHNSON J H. Some structures and notation of Q-analysis [J]. Environment and Planning B, 1981, 8: 73.
[8] ATKIN R H. An algebra of patterns on a complex I [J]. Int. J. Man-Machine Studies, 1974, 6: 285.

[9] ATKIN R H. An algebra of patterns on a complex II [J]. Int. J. Man-Machine Studies, 1974, 8: 483.

[10] JOHNSON J H. Hypernetworks in the science of complex systems [M]. London: Imperial College Press, 2013.

[11] MALETIĆ S, RAJKOVIĆ M. Combinatorial Laplacian and entropy of simplicial complexes associated with complex networks [J]. Eur. Phys. J. Special Topics, 2012, 212: 77.

[12] DEGTIAREV K Y. Systems analysis: Mathematical modeling and approach to strucutral complexity measure using polyhedral dynamics approach [J]. Complexity International, 2000, 7: 1.

[13] CASTI J L.Alternate realities: Mathematical models of nature and man [M]. USA: John Wiley & Sons, 1989.

[14] GOULD P, JOHNSON J, CHAPMAN G. The structure of television [M]. London: Pion Limited, 1984.

[15] MALETIĆ S, RAJKOVIĆ M, VASILJEVIĆ D. Simplicial complexes of networks and their statistical properties [J]. Lecture Notes in Computational Science, 2008, 5102(II): 568–575.

[16] GHRIST R. Barcodes: The persistent topology of data [J]. Bull. Amer. Math. Soc. (N.S.), 2008, 45(1): 61.

[17] CARLSSON G. Topology and data [J]. American Mathematical Society Bulletin, New Series, 2009, 46: 255.

[18] EDELSBRUNNER H, HARER J L. Computational topology: An introduction [M]. Providence: American Mathematical Society, 2010.

[19] EDELSBRUNNER H, HARER J L. Persistent homology — a survey [J]. In Surveys on discrete and computational geometry, 2008, 453: 257–282.

[20] EDELSBRUNNER H, LETSCHER D, ZOMORODIAN A. Topological persistence and simplification [J]. Discrete and Computational Geometry, 2002, 28: 511.

[21] ZOMORODIAN A, CARLSSON G. Computing persistent homology [J]. Discrete and Computational Geometry, 2005, 33(2): 249.

[22] GOLDBERG T E. Combinatorial Laplacians of simplicial complexes [M]. New York: Annandale-on-Hudson, 2002.

[23] DUVAL A M, REINER V. Shifted simplicial complexes are Laplacian integral [J]. Transactions of the American Mathematical Society, 2002, 354(11): 4313.

[24] MALETIĆ S, HORAK D, RAJKOVIĆ M. Cooperation, conflict and higher-order structures of complex networks [J]. Advances in Complex Systems, 2012, 15: 1250055.

[25] NEWMAN M E J. Networks: An introduction [M]. Oxford: Oxford University Press, 2010.

[26] MOHAR B. The Laplacian spectrum of graphs [J]. Graph Theory, Combinatorics, and Applications, 1991, 2: 871.

[27] HODGE W V D. The theory and applications of harmonic integrals [M]. Cambridge: Cambridge at the University Press, 1952.

[28] FRIEDMAN J. Computing Betti numbers via combinatorial Laplacians [C]// Proceedings of the Twenty-Eighth Annual ACM Symposium on the Theory of Computing, Philadelphia, Pennsylvania, USA, May 22–24, 1996, page number 386.

[29] CHUNG F R K. Spectral graph theory [M]. USA: American Mathematical Society, 1996.

[30] BANERJEE A, JOST J. Graph spectra as a systematic tool in computational biology [J]. Discrete Applied Mathematics, 2009, 157: 2425.

3

How Do We Build Simplicial Complexes

In previous chapter we scratched the surface of large potential of simplicial complexes as mathematical objects. Just like different approaches of simplicial complexes' analysis exist, we can foresee that the applicability of simplicial complexes and algebraic topology is versatile. At the first sight, the content of the previous chapter may seem abstract and rather confusing, hence now is the time to introduce a bridge between mathematical concepts and tools, and real-world applications. Of course, the complete list of applicability is not exhausted, but the choice of methods of simplicial complex for mation are made to cover as broad applicability as possible, and to inspire new paths of the research.

In this chapter we will not focus on calculations of algebraic topology quantities of built simplicial complexes, but leave it for the upcoming chapters. From the Introduction we can notice the omnipresence of complex networks everywhere around us, and their properties which emerge from their mesoscopic structures, which originate between global and local structures, draw a particular attention. It turned out, for example, that simplicial complexes are convenient for capturing such mesoscopic structures.[1–3] Hence, first we will introduce different ways of building simplicial complexes from graphs, i.e., complex networks. Next we will deal with the ways of building a simplicial complexes from datasets embedded in some metric space, and finally we will introduce building simplicial complexes from time series. The order of sections is not arbitrary, since some of the methods defined in one section can be used in the next.

3.1   From complex networks (graphs)

The simplest way to represent mathematically a system formed by elements and their pairwise interactions are graphs, and as mentioned in the Introduction such system is called complex network. Mathematically a graph is defined as a set of N vertices (or nodes) connected by links (or edges).[4, 5] Links can be either directed, if a direction is specified along them, or undirected, if no direction is specified. Accordingly, the whole graph is denoted as directed (Fig. 3.1(a)) or undirected (Fig. 3.1(b)). More precisely, undirected links are rather bidirectional ones, since they allow crossing in both directions. For this reason an undirected graph can always be thought of as a directed one where each undirected link is replaced by two directed ones pointing in opposite directions. Even with the visual inspection of Fig. 3.1 we can notice that links define the neighborhoodness between nodes, and their number is important for the characterization of a node. Namely, *the degree* of a node is a basic property characterizing a single node, measuring its number of neighbors. Hence, in Fig. 3.1(b), for example, the node 1 is connected to nodes 2, 3, and 4, so we say that node 1 has degree equal to 3, and in a similar way we can count the number of neighbors of other nodes, resulting in degrees 4, 2, 2, and 1, for nodes 2, 3, 4, and 5, respectively. A related quantity is the degree distribution $P(k)$, which is the probability that a randomly chosen node has degree k.

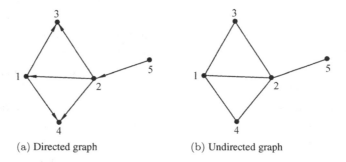

(a) Directed graph   (b) Undirected graph

Fig. 3.1   Examples of directed, and undirected graphs

Simplicial complex from directed graph

From the Fig. 3.2(a) we can notice that some nodes have a role of source and some have a role of sinks, and some have the both roles. For example, nodes 1, 2, 5 are sources, nodes 1, 2, 3, 4 are sinks, whereas nodes 1, and 2 are the both. Let us form two sets: the *out*-set X which contains only the source vertices, and the *in*-set Y which contains only the sink vertices. Obviously, due to the dual role of some vertices, these two sets may overlap. If we recall the definition of a simplicial complex from relation, we are half a way in building a simplicial complex from directed graph, since only remains to define a binary relation. This task is simple as well since binary relation naturally emerges from the pairwise connections between nodes. Hence, we can define a relation between two sets as "an element x from the set X is related to an element y from the set Y if between them exists a link, and an element x is a source, whereas an element y is a sink". In other words, a simplicial complex from directed graph is built by vertices which are sink nodes in the original graph, and simplices which are associated to the source nodes in the original graph.[6] A careful reader may notice that alongside with this simplicial complex, we can build a conjugate complex, with reversed roles: vertices are the source nodes, whereas simplices are the associated to the sink nodes. Take, for example, directed graph from the Fig. 3.2(a) elements of the set X are $\{1, 2, 5\}$, and the elements of the set Y are

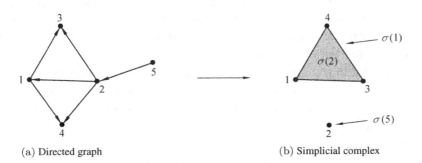

(a) Directed graph   (b) Simplicial complex

Fig. 3.2   Building a simplicial complex from a directed graph

$\{1, 2, 3, 4\}$, hence the simplices, illustrated in Fig. 3.2(b) are:

$$\sigma_1(1) = \langle 3, 4 \rangle$$
$$\sigma_2(2) = \langle 1, 3, 4 \rangle$$
$$\sigma_0(5) = \langle 2 \rangle$$

where subscripts are associated to the dimension of a simplex, forming 2-dimensional simplicial complex. Although two sets overlap the roles of the elements of each set are different. As a simple exercise we suggest to the reader write explicitly the simplices of conjugate complex, and to draw the associated geometrical representation.

Clearly, by building two simplicial complexes from a single graph we obtained the information about the relationship between graph's elements which transcends the information obtained solely from the pairwise relations. These advantages will be even more supported very soon when we introduce the other ways of building a simplicial complex from graph, i.e., complex network.

Clique complex

From examples in Fig. 3.1 we can notice that there are some groups of nodes in which every node is connected to every other node. For a current purposes we will focus only on Fig. 3.1(b). Originating from social science (specifically social network analysis), such groups of nodes are called *cliques* in analogy with the highly connected social groups. Pursuing the same analogy we may wonder whether there is a way to capture the relationship between different cliques with respect to the nodes which are members of them, and reversely, we may wonder whether there is a way to capture the relationship between different nodes with respect to the cliques of which they are members. These ideas give us already a hint: form two sets, one which contains all maximal cliques X, and one which contains all vertices Y. Then introduction of a relation is simply induced by the node's membership to the clique. In this way we are provided with all necessary building blocks to form a simplicial complex. Namely, a simplicial complex in which simplices are all maximal cliques and vertices are all nodes

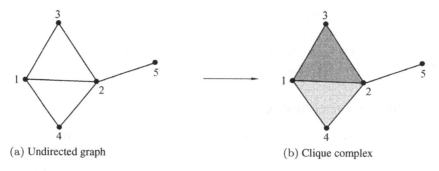

(a) Undirected graph   (b) Clique complex

Fig. 3.3   Building a clique complex from an undirected graph

of the initial graph is called the *clique complex*.[7,8] Formation of a
simplicial complex by reversing the roles of cliques and vertices, i.e.,
cliques play the role of vertices, and simplices are associated to the
nodes of the original graph, is called the *conjugate clique complex*.
Intuitively, the meaning of the conjugate clique complex is simple: we
want to extract the information about the relationship between nodes
with respect to the cliques to which they belong. The advantage of
this approach with respect to the graph approach is even more clear
in this case than in the case of directed graph, described above.

Finally, let us clarify the above definitions by an example. From
Fig. 3.3(a) we find 3 maximal cliques (that is cliques which are
not part of other cliques) depicted in different colors in Fig. 3.3(b):
$\{1, 2, 3\}$, $\{1, 2, 4\}$, and $\{2, 5\}$ and let us label cliques a, b, and c,
respectively. Hence, the set $\{a, b, c\}$ is our clique-set, and together
with the vertex set $Y = \{1, 2, 3, 4, 5\}$ build a 2-dimensional simplicial complex with simplices:

$$\sigma_2(a) = \langle 1, 2, 3 \rangle$$

$$\sigma_2(b) = \langle 1, 2, 4 \rangle$$

$$\sigma_1(c) = \langle 2, 5 \rangle$$

where subscripts are associated to the dimension of a simplex. As a
simple exercise we suggest to the reader to write explicitly the simplices of conjugate complex, and to draw the associated geometrical

representation. The interpretation of calculation of topological quantities on a conjugate clique complex can be even more interesting and important than for a clique complex. For easy understanding, the different colors in Fig. 3.3 indicate maximal clique simplices.

Neighborhood complex

We have noticed that in building a simplicial complex from directed graph two sets may overlap, nevertheless in the case of an undirected graph they are identical. Namely, from an undirected graph we can build a simplicial complex called the *neighborhood complex*[1, 9–11] with the same set of vertices, and to each node i from the original graph is associated a simplex $\sigma(i)$ in the neighborhood complex, defined by vertices which are linked to i in the original graph. Or in other words, the simplices are all subsets of the vertex set of the underlying graph that have a common neighbor. This definition seems a bit confusing, but actually it is just a generalization of building a simplicial complex from directed graph. Of course, since this simplicial complex is built on one set, the neighborhood complex and its conjugate are the same. Take an example in Fig. 3.4 where to each node is associated a set of its neighbors, that is $\{2, 3, 4\}$, $\{1, 3, 4, 5\}$, $\{1, 2\}$, $\{1, 2\}$, $\{2\}$, and hence to each node is associated a simplex, that is $\sigma_2(1), \sigma_3(2), \sigma_1(3), \sigma_1(4), \sigma_0(5)$, respectively. In Fig. 3.4(b) we illustrated the geometrical representation of neighborhood complex,

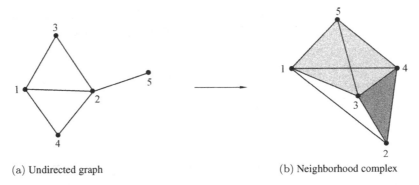

(a) Undirected graph   (b) Neighborhood complex

Fig. 3.4   Building a neighborhood complex from an undirected graph

with simplices depicted in different colors. Note that simplices $\sigma_1(3)$ and $\sigma_1(4)$ overlap. It is easy to check that two simplices are q-near if their associated nodes in the underlying graph have $q+1$ common vertices. We suggest to the reader to check other similarities, like determination of the relation between simplex dimension, and the degree of associated node in the underlying graph. In Fig. 3.4 the colors emphasize matching between nodes of the original graph and associated simplices in the neighborhood complex.

At the first sight for practical real-world applications, building neighborhood complex seems unnecessary since we are making already simple situation more complicated. Nevertheless, it does not have to be the case, since the analysis of such simplicial complex may reveal hidden patterns of relationships between the elements of complex network.

Independence complex

Links between nodes in complex network obviously carry a natural information about the existence of relationship between two nodes, and the collection of nodes where every node is connected to every other node carries an information about the dense aggregations of nodes. Such structures are building blocks of clique complex and its conjugate from the underlying complex network. But the nonexistence of links between nodes may be quite important as well. In other words, for some particular real-world complex networks it may be of a great importance to examine the missing relationships between nodes. Hence, for some graph G we first build the complement graph of G, that is, a graph in which two nodes are connected if they are not connected in graph G, and vice versa, two nodes are disconnected in complement graph if they are connected in graph G. By finding all maximal cliques of this new graph (called independence sets, or anticliques) we are building the so called independence complex.[7] To make a relationship with the underlying graph, let us emphasis that the vertices of independence complex are nodes of the underlying graph G, whereas simplices are maximal cliques (together with all their subcliques) of the complement graph of G.

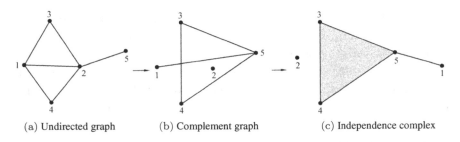

(a) Undirected graph   (b) Complement graph   (c) Independence complex

Fig. 3.5   Building an independence complex from an undirected graph

Figure 3.5 illustrates the way of construction of an independence complex starting from an arbitrary graph (Fig. 3.5(a)). At the first stage (Fig. 3.5(b)) we build a complement graph with the same set of nodes but the new set of links $\{(1,5),(3,4),(3,5),(4,5)\}$, and we see that the node 2 is isolated. Finding all maximal cliques $\{1,5\}$, $\{2\}$, $\{3,4,5\}$ of such a graph, and labeling them like a, b, c, respectively, builds an independence complex illustrated in Fig. 3.5(c), with simplices depicted in different colors:

$$\sigma_1(a) = \langle 1,5 \rangle$$
$$\sigma_0(b) = \langle 2 \rangle$$
$$\sigma_2(c) = \langle 3,4,5 \rangle$$

where subscripts are associated to the dimension of a simplex. Like in the cases of building a simplicial complex from directed graph or from cliques, with an independence complex we can associate the conjugate complex, again leaving to the reader to explicitly write and draw appropriate simplices. In Fig. 3.5 different colors indicate maximal cliques, i.e., simplices, of complementary graph.

Matching complex

We have seen so far that working with simplicial complexes things can get very complicated, but at the bottom line the intuition that lies behind it can be relatively simple. Let us focus now on links, which are usually labeled as (un)ordered pairs of nodes. Hence, if we would like to examine the relationships between links, it would be

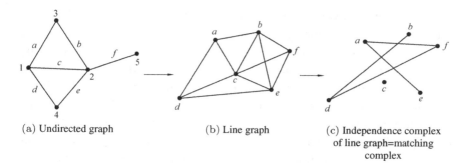

(a) Undirected graph   (b) Line graph   (c) Independence complex
                                                          of line graph=matching
                                                          complex

Fig. 3.6   Building a matching complex from an undirected graph

convenient to transform graph in some suitable way. First we build
a *line graph* in which nodes are associated to links of the underlying
graph, and two nodes in line graph are connected if their associated links in the underlying graph share a node. The independence
complex of line graph is called the matching complex.[12] In other
words, the vertices of matching complex are the edges of the underlying graph G and simplices are sets of edges of G with no two edges
having a common vertex, that is, the matching complex is a clique
complex of the complement graph of the line graph of G. Observing
Fig. 3.6 will help to clarify these rather confusing definitions. Starting from an arbitrary graph, let us first label all edges (1,3), (2,3),
(1,2), (1,4), (2,4), (2,5), like a, b, c, d, e, f, respectively (Fig. 3.6(a)).
Then we build a line graph with nodes $\{a, b, c, d, e, f\}$, connected
by edges $(a, b), (a, c), (a, d), (b, c), (b, e), (b, f), (c, d), (c, e), (c, f), (d, e),$
(e, f) (see Fig. 3.6(b)). Next we build a complement graph of line
graph with the same set of vertices, but with edges: $(a, e), (a, f),$
$(b, d), (d, f)$, the vertex c isolated. Finally, we find all maximal cliques
of that graph and build (in this case) 1-dimensional matching complex, illustrated in Fig. 3.6(c) with simplices depicted in different
colors. In Fig. 3.6 different colors indicate simplices in the independence complex of the line graph.

Notes

The above list of different building ways of a simplicial complex
from complex network does not exhaust all possibilities, but gives

an introductory overview of the richness in simplicial complex construction. In these short notes we want to emphasis some important points. First, from the illustrative examples in the above definitions we can see that from the same graph we can build different simplicial complexes. This means that from a single graph we can build a clique complex, neighborhood complex, independence complex and matching complex, and all of them can be analyzed by the same apparatus emerging from algebraic topology. For example, Q-analysis or homology or combinatorial Laplacian can be calculated for each simplicial complex giving rise to the abundance of information about the underlying graph, which we could not get from the graph-theoretic analysis solely. Second, simplices built on complex networks, and their collections revealed through q-connectivity classes, can be used for the examination of mesoscopic structures, called *communities*,[4] which emerge in real-world complex networks, and in Ref. [3] *q-dimensional simplicial communities* are introduced as the aggregations of simplices at various dimensions, namely the q-levels. Hence, since there are different simplicial complex representations of the complex network, different simplicial communities emerge.

3.2 From data in metric space

For building a simplicial complex from graphs the distances between nodes were not important, what mattered was whether some kind of relationship between nodes exists. For example, modeling the relationship between two web-pages through the hyperlink does not require the knowledge of physical distance between the host servers of two web-pages, or for modeling a group of persons who participate in a meeting it is not important to know how far persons sit from each other. Nevertheless, we can deal with the data sets for which the distances between elements (i.e., points or vertices or nodes) are of essential importance, like in the system of sensors or the system of mobile phone towers, where the relationship between two elements depends on the overlapping of coverage of each sensor or tower. For such systems, or generally data sets, the elements are associated with coordinates of a Euclidean metric space, hence we may say that the data are of geometric nature. What we mean by Euclidean metric space that it is a set of points for which between all elements of a set an Euclidean distance is defined, and recall that Euclidean distance between two points $p_1 = (x_1, y_1, z_1)$ and $p_2 = (x_2, y_2, z_2)$ in some 3-dimensional space is defined like

$$d(p_1, p_2) = \sqrt{(x_1 - x_2)^2 + (y_1 - y_2)^2 + (z_1 - z_2)^2}.$$

Hence, having a collection of data points with the above properties we may be interested in reconstruction of the global relational structure which originate from geometrical distances. Or in other words, we want to reconstruct the shape of data, and use the homology, and eventually Q-analysis, to describe data. More precisely, suppose that we are provided with the data points distributed in 2-dimensional Euclidean space as presented in Fig. 3.7, and we are wondering what kind of simplicial complexes can be built on such data set. Just as in the case of building a simplicial complexes from a graph, the current case displays a different opportunities as well. As we will see, some methods of building a simplicial complex from a data set may be related to some of the methods of building a simplicial complex from a graph. Of course, like in the previous section, the introduced methods does not exhaust all possibilities, but try to cover the most commonly used.

Before we address the issue of building different simplicial complexes from data points we will present a short intuitive justification

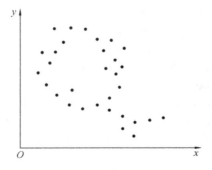

Fig. 3.7   An example of data points distributed in 2-dimensional Euclidean space

for building such complexes. Say, in some space we have a set of open balls $S = \{S_1, S_2, \ldots, S_N\}$, such that an aggregation of balls and their intersections is contractible, and their union $\cup S_i$ is the space of interest. If we make each S_i a vertex and build a q-simplex whenever the intersection $S_{i0} \cap S_{i1} \cap \cdots \cap S_{iq} \neq \emptyset$ appears, then actually we build an abstract simplicial complex $N(S)$. This technique originates from the topology and it is called the nerve of the cover.[13, 14] Hence, put it intuitively, the collection of balls covers the underlying space of interest. In order to relate the aggregation of balls and built simplicial complex, the Nerve lemma states that simplicial complex $N(S)$ and $\cup S_i$ has the same homotopy type.[13, 14] Although we did not define homotopy, since it is out of the scope of this book, for a reader is sufficient to know that when two topological spaces have the same homotopy type, then they cannot be distinguished homologically. For more thorough introduction to homotopy, as well as to the Nerve lemma, can be found elsewhere (say [13] or [14]).

Čech complex

Having the set of data points V in some metric space X, and some parameter $r \in \mathbf{R}$, we can take data points as centers of balls of radius r, that is $B(v_i, r)$, for each $v_i \in V$

$$B(v_i, r) = \{x \in X | d(v_i, x) < r\}$$

whenever two balls intersect, we build an edge (i.e., 1-simplex) between two vertices centered at those balls, when three balls have nonempty intersection, we build a triangle (i.e., 2-simplex) by three vertices centered at those balls, and so on. Analogously, following the same way of construction, we can build the higher-dimensional simplicial complex, called the Čech complex[7] in which the data points take the role of the vertex set. Clearly, the construction of Čech complex is associated to the nerve of the cover, and consequently satisfy the Nerve lemma, recovering in this way the shape of space built by data points. The careful reader probably noticed that the arbitrariness of radius' value does not guarantee that we will accurately reconstruct the shape of the data, and hence some

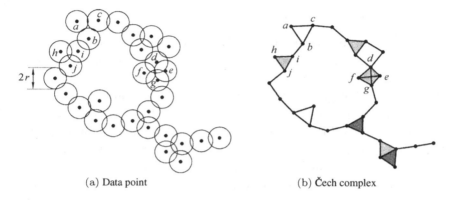

(a) Data point   (b) Čech complex

Fig. 3.8   An example of building the Čech complex from data points

additional procedure is necessary for tackling this problem. This procedure lies, actually, in the persistent homology filtration which can be conducted by tuning the parameter, and building a series of nested Čech complexes[15] for different increasing values of radius, $r_1 < r_2 < r_3 \cdots$. For a moment we will skip an introduction of the persistent homology of Čech complex, and return to it in the following chapter which is focused on the applications.

For the current purposes of defining the Čech complex, let us take an example of 2-dimensional data set from Fig. 3.7, and around each point draw a circle of an arbitrary radius r (Fig. 3.8(a)). In Fig. 3.8(b) circles are omitted, whereas the Čech complex which emerges from the overlapping circles is drawn, and the higher-dimensional simplices are indicated by different shades of gray. If we pay attention on some particular subsets of points we will see how it is easy to interpret the shape of data. For example, a subset of points $\{a, b, c\}$ builds 1-dimensional hole, since the circles overlap only pairwise, whereas subset of points $\{h, i, j\}$ builds a triangle, that is, 2-dimensional simplex, since all three circles around these points have common intersections. The same holds for four points $\{d, e, f, g\}$ which build 3-dimensional simplex. With further expansion of the radius 1-dimensional holes may be filled and only the central large hole will survive.

Vietoris-Rips complex

Similarly to the construction of Čech complex we can build yet
another simplicial complex from the intersection of balls centered
around the data points, but using the different criteria. Namely,
instead of looking for whether three, four, and so on, balls intersect,
and then build the higher-dimensional simplices directly, we may
restrict computation on the emergence of only pairwise overlappings
between balls, and then build simplices. Whenever two balls intersect,
we connect by an edge two data points located at the centers of those
balls. Clearly, the structure that emerge is the 1-dimensional simplicial complex, that is, the graph, and by building a clique complex
from this graph we obtain the so called the Vietoris-Rips complex.[7]
Unlike the Čech complex, in principle the Vietoris-Rips complex does
not satisfy the conditions of the Nerve lemma, but with the proper
choice of radius which is related to Čech-radii, the conditions may
be satisfied. In some sense, the Vietoris-Rips complex may be understood as a special case of clique complex, for which the underlying
graph is constructed by the edges that are not longer than $2r$. Like
in the case of the Čech complex, the difficulty to find correct choice
of value of r for approximation of the shape of data impose the application of persistent homology filtration by tuning the parameter r,
like $r_1 < r_2 < r_3 \cdots$.

The Fig. 3.9 illustrates building the Vietoris-Rips complex from
the same set of data as for the Čech complex, and the higher-
dimensional simplices are indicated by different shades of gray.
Comparing Fig. 3.8(b) and Fig. 3.9(b) we can notice significant
differences in the number of higher-dimensional simplices. Namely,
whereas in the case of the Čech complex the subset of points $\{a, b, c\}$
build a hole, in the case of the Vietoris-Rips complex they build the
2-dimensional simplex, and with a simple visual inspection of both
figures we can see that yet one more 2-dimensional simplex emerges
in the Vietoris-Rips complex. The other higher-dimensional simplices
from the Čech complex appear in the Vietoris-Rips complex as well,
since six edges (i.e., 1-dimensional simplices) which build a clique
in the Vietoris-Rips complex are actually six 1-dimensional faces of

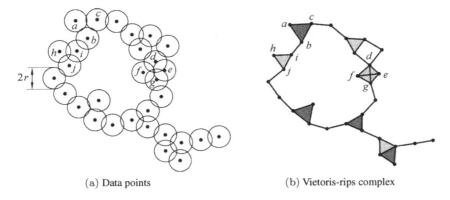

(a) Data points   (b) Vietoris-rips complex

Fig. 3.9   An example of building the Vietoris-Rips complex from data points

3-simplex in the Čech complex. Note that the radius is the same in Fig. 3.8(b) and Fig. 3.9(b), and for a suitable choice of different radii the Vietoris-Rips complex may recover the same structure as the Čech complex,[16] that is to satisfy the Nerve lemma.

Witness complex

When we are dealing with large data sets the manipulation with the Vietoris-Rips complex, as well as the Čech complex, may be impractical and computationally very demanding. To overcome this obstacle we can resort to the solution based on building a simplicial complex which will result in reducing the size of data set. If we have a large set of data points X in some Euclidean space, we can choose a small set of points $L \subseteq X$ called the *landmarks*, and together with the set of points $X \backslash L$, called the *witnesses*, we build a simplicial complex on this set of points instead of X in the following way. A q-simplex $\sigma_q = \langle l_0, l_1, \ldots, l_q \rangle$ is weakly witnessed by $x \in X \backslash L$ if $d(l, x) \leq d(k, x)$ for every $l \in \{l_0, l_1, \ldots, l_q\}$ and $k \in L \backslash \{l_0, l_1, \ldots, l_q\}$. In other words, a subset of points $\{l_0, l_1, \ldots, l_q\} \in L$ is a simplex if and only if there is a point (a witness) $x \in X \backslash L$ with every point in $\{l_0, l_1, \ldots, l_q\}$ closer to x than to any other point in $L \backslash \{l_0, l_1, \ldots, l_q\}$. Finally, the witness complex is defined as the collection of all simplices and their

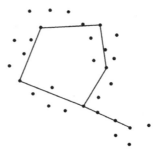

Fig. 3.10   An example of building the witness complex from data points

faces built on the vertex set L that are weakly witnessed by a points in X.[17]

The above definition of witness complex can easily be related to the way of building a simplicial complex from relation between two sets, in our case: the landmark set and the witness set.

Although the construction of a witness complex reduces the size of the sample, the challenge lies in the convenient choice of the landmark set in order to preserve the topology of the original data set. Like in the cases of the Vietoris-Rips complex and the Čech complex where the indeterminacy of the radius led to the conveniences of the persistent homology calculations, the same holds for the witness complex.[18] We will not go into details of defining the best way for choosing the landmark set, but we will illustrate by an example the way of building the witness complex. In Fig. 3.10 our sample is again a set of 31 data points from Fig. 3.7, and we have chosen randomly 6 landmark points, which build 1-dimensional simplicial complex. Furthermore, this simplicial complex contains large 1-dimensional hole, the information which would be recovered even if we connect all points mutually. Obviously, this coarse-graining procedure reduced the number of data points, as well as the number of simplices, while the homological structure is preserved.

Notes

If we are interested on homological characterization of the shape of data, then building Čech or Vietoris-Rips complexes has a drawback, since the higher-dimensional simplices may occur. Namely, these higher-dimensional simplices does not participate in homology group calculations, and in this case may carry an unnecessary computational burden. Hence building the witness complex may be more acceptable choice. Nevertheless, if we are interested in the Q-analysis calculations, the higher-dimensional simplices are of an essential importance.

It is important to point that the definition of witness complex, as it is introduced here, is usually used for defining the so called *weak witness complex*, and additionally the *strong witness complex* can be defined.

Examples drawn in Figs. 3.8–3.10 may lead to the wrong conclusion that it is not important which simplicial complex we build from data points since all may preserve the global topological characteristics (i.e., homology), we want to stress that it does not have to be the case. Namely, like the number of data points, their particular distribution in space, and many other things, will influence the right choice of simplicial complex which we want to build.

3.3 From time series

In some sense this section can be understood as an application of the previous sections, and it is based mainly on the research presented in [19]. Although there are the other ways of building simplicial complexes from time series, we will try to avoid whenever possible, and if it is not necessary, confusion for the reader. Hence, the topics of previous two sections are well developed and find an application in many research fields, whereas the content of the present section is partially in a developing phase. Namely, the extraction of properties that govern the dynamics of complex system can be revealed from geometrical and topological features of phase space and it is well known for some time, nevertheless, the interest in relating them with simplicial complexes draw larger attention only recently.

In plain words, often the only information that we have about the dynamical complex system is time series, i.e., the time sequence of measurements of one variable, and what we do not know is the exact number of variables which govern the dynamics of a complex system. The question emerges: how can we extract from a time series relevant information about the dynamical system which generate it? Hence we need to devise methods for determination of the number of variables, as well as methods which reveal the relationships between them, that will contribute to the full description of the underlying complex system. The procedure which merges these two tasks is included in the phase space reconstruction, that is, the reconstruction of the space of all states, in which coordinates of one point are associated to one state of the system. Clearly, the dimension of phase space, and hence the number of coordinates that characterize a point in phase space, is equal to the number of variables. One of the aim of the research in time-series analysis is to get an information about the dynamics behind some observed time-ordered data.[20] In the current introduction we will work only with deterministic systems, that is systems for which when its present state is fixed, the future states are determined as well.

The task of, as well as the introduction in, phase space reconstruction is not trivial, and requires defining concepts which are out of the scope of this book. Nevertheless, since our intention is to give a broad introduction in the applicability of simplicial complexes, we will introduce only necessary definitions and concepts, and we will try to be as much as possibly consistent in our exposition.

Phase space reconstruction

The time series is often the first and only experimental data about the behavior of a complex system. Although the knowledge of all components of phase space for the complete determination of system's characteristics is often lacking, for the determination of some characteristics the knowledge of time changes of only one variable is enough. Generally, for reconstructing a d-dimensional phase space, from a single time series, one usually takes samples with time delay, and equates the reconstructed phase space point with d consecutive points of the time series separated by a time delay. Namely, usually one has an observed scalar time series, say $X = \{x_1, x_2, \ldots, x_n\}$ of some variable X measured at even time intervals. Takens' embedding

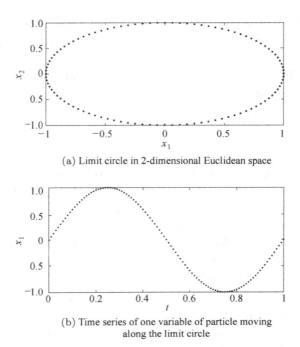

(a) Limit circle in 2-dimensional Euclidean space

(b) Time series of one variable of particle moving
along the limit circle

Fig. 3.11   An example of limit circle

theorem[21] states that we can reconstruct a phase space from time series considering time delays, so that the data is lifted to d dimensions using time-delay embedding. In that way the data becomes a path in a higher-dimensional phase space of unknown dimension and shape. The points in the reconstructed space usually converge to a manifold or some other subspace of the n-dimensional Euclidean space or attractors of fractal dimension.[22]

Take for example a movement of a particle along the limit circle (Fig. 3.11(a)). Moving along the limit circle is characterized by two variables x_1 and x_2, and equations of motion $x_1 = \sin(2\pi t)$ and $x_2 = \cos(2\pi t)$. Then movement of particle can be written like

$$x(t) = (x_1, x_2) = (\sin(2\pi t), \cos(2\pi t)) = \left(\sin(2\pi t), \sin\left(\frac{2\pi t + \pi}{2} \right) \right)$$

$$= \left(x_1(t), x_1\left(\frac{t+1}{4} \right) \right)$$

Hence, movement along the limit circle can be represented by only one variable (see time series in Fig. 3.11(b)), but with the parameter shifted for $1/4$, i.e., the time delay.

Practically, the procedure is following. Take the m uniformly spaced samples of the time series of the observed variable and concatenate them into a single vector,

$$v_i(t) = [x_t, x_{t-\tau}, x_{t-2\tau}, \cdots, x_{t-(m-1)\tau}] \tag{3.1}$$

and in that way we are mapping the state vector of a dynamical system to a point in the reconstructed space. Hence, we have formed m-dimensional vectors. The formation of vector (3.1) is followed by the arbitrary choice of two parameters, which are therefore of crucial importance for the phase space reconstruction procedure. These two parameters are τ, the time delay, and m, the embedding dimension. Takens' embedding theorem guarantees the preservation of topological properties of the attractor however not of its geometrical properties. Consequently, the choice of the time delay and the embedding dimension has a large influence on the accuracy of the obtained results in applications. The embedding theorem[21] requires that the embedding dimension should satisfy $m > 2d$, where d is the real dimension of the dynamical system, that is, the real number of variables. Usually, the first step in the reconstruction procedure is the estimation of the (optimal) time delay τ, and once this goal has been achieved the appropriate embedding dimension m is determined. The large number of techniques for estimations of τ and m gives rise to the nontrivial aspect of phase space reconstruction. Two most often used methods for estimation of the delay time are based on determination of either the first minimum of the average mutual information[23] or the first zero value of autocorrelation function.[24] For the purpose of this short introduction, applications of simplicial complexes in time series analysis, we will use the first local minimum of the mutual information. We will omit the details of calculation of the mutual information, as it is out of the scope of the book, and since our focus is on the concrete applications of simplicial complexes. The reader interested in this topic can find the details in referenced research papers.

The next step is the estimation of embedding dimension. At this point we will stop with the conventional approach, and start applying the tools from algebraic topology that we introduced so far.

Persistent homology of phase space

Once that the time delay is determined, we can change values of embedding dimension, that is for each value $m = 2, 3, 4, \ldots$ we build a simplicial complex, and it captures the topology of the reconstruction space followed by the calculation of topological invariants and properties using methods of the topological data analysis.[25] Maybe this sounds little blurry, so we will see what have we done so far. From single variable time series we built m-dimensional vectors, which, in other words, are the coordinates of points in m-dimensional space. Since we want to recover the topological structure of the set of points in m-dimensional space, we can apply some of the ways of building a simplicial complex from the data set embedded in metric space, which we introduced in the previous section, say the Čech complex. But the shape of the Čech complex depends on the radius of balls around data points, and the choice of radius is arbitrary, unless we devised some criteria for estimation of convenient radius values. Hence, in the absence of criteria, the radius serves as a free parameter and many different Čech complexes can be built. Nevertheless, if we jump to the previous chapter, where we introduced different ways of characterization of simplicial complexes, we can recall that we are already equipped with the well-developed procedure to build and characterize simplicial complexes: persistent homology.

Take for example a simple pendulum. From time series we build a 2-dimensional vectors associated to the data points in 2-dimensional embedding space. It is known that we have to reconstruct the limit cycle, much like the one in Fig. 3.11. In Fig. 3.12(a)–(f) we illustrate building the Čech complex from data points distributed along the closed orbit. The Čech radius in Fig. 3.12(a) is small, hence the structure is disconnected ($\beta_0 > 1$), and there is only one pair of points which forms a 1-dimensional simplex. By increasing the Čech radius, in Fig. 3.12(b) new simplices emerge, including higher-dimensional,

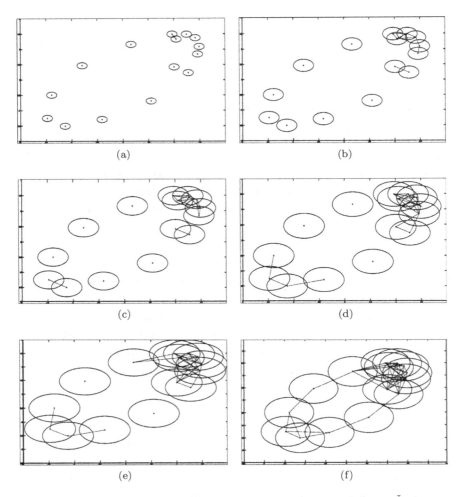

Fig. 3.12 The stages of persistent filtration through building different Čech complexes by increasing radius around each data point

like 2-simplices. The further increase of radius (see Fig. 3.12(c)–(e)) leads to the emergence of more and more simplices, and the decrease in the connected components, and eventually ends with one connected component (Fig. 3.12(f)). At the last stage we can see that alongside with the existence of one connected component, the simplicial complex has one 1-dimensional hole, that is $\beta_0 = 1$, which is the topological characteristic of a limit circle. In this way, the computation of the persistent homology is performed following the construction of the filtered Čech complex, and following the simple procedure we have reconstructed the topological shape of pendulum's phase space.

Instead of forming 2-dimensional vectors and associating them with the points in 2-dimensional embedding space, we also formed 3- or 4-dimensional vectors and applied the same procedure, the same topological structure would be recovered. That means that in higher dimensions the robust topological structure (like 1-dimensional holes) is preserved. Since it is not easy to draw simplicial structures in higher dimensions, we will omit it. In the following chapter we will apply these methods on some concrete examples, and the real potential of applicability will be clearer.

Notes

Instead of building the Čech complex over the data set, we could use either the Vietoris-Rips complex or the witness complex, and apply persistent homology filtration. Nevertheless, using any of these methods has some advantages and disadvantages. For example, for the Čech complex we know that it satisfies the Nerve lemma which provide us the certainty that we will recover the topological structure. On the other hand, the Vietoris-Rips complex or the witness complex can be computationally less demanding. For example, the advantage of building the witness complex from data points is that it can be conveniently used for coarse-graining the data,[18] in the sense of reducing the number of data points, and at the same time, under some conditions, preserve the topology that underlies the data set.

3.4 Summary

Previous chapter was devoted to the introduction of basic concepts of simplicial complexes, and the quantities which characterize them. But for the applications of the machinery of algebraic topology it is not enough. We need to know how we can relate simplicial complexes to the particular research framework, or in other words, how can we build a convenient simplicial complex from some real data and apply the tools from algebraic topology. That task led us to the current chapter and the short overview of possible ways of building a simplicial complexes from data.

We have considered three types of data from which we can build simplicial complexes, in attempt to cover as much as possible broad applications. Complex networks are omnipresent everywhere, and hence it is not surprising that so much research attention is devoted to them. The graph theory emerges as the natural setting for mathematical representation of complex network, and therefore the applications of tools from the graph theory are overwhelming. In order to uncover the higher-order structures, and the relationships between them, which cannot be computed with graph-theoretical tools, simplicial complexes are found as extremely suitable. Hence, we introduced five ways of building a simplicial complex from a graph, each one capturing different structural properties. The straightforward construction of a simplicial complex from a graph include building clique complex or neighborhood complex, where in the former links of the original graph are preserved and cliques (i.e., complete graphs) take the role of simplices, and in the latter the simplicial complex is built over group of nodes on the basis of adjacency to the common node. The construction of the independence complex is related to emergence of structures built from the missing links, whereas the construction of the matching complex captures the structural relationships between missing links in a graph.

When we are dealing with some distribution of data points in some metric space, like in computer graphics, it is convenient and often necessary to recover the topological structure which is carried by these points. Although there are many different methods, each one with advantages and disadvantages, we have introduced three: the Čech complex, the Vietoris-Rips complex and the witness complex. Satisfying the conditions of the Nerve lemma, building the Čech complex over data points in metric space completely recovers topological structure, whereas for the Vietoris-Rips complex and the witness complex it is satisfied under certain conditions.

Finally, we have merged the methods of persistent homology and the Čech complex in reconstructing the phase space from a scalar time series. Namely, the outcome of an experiment carried over some dynamical system is often a time series of numbers, i.e., measurements, and the knowledge of the real number of variables responsible for system's dynamics is lacking. In order to reconstruct the phase space in which coordinates of each point represent the state of system, different methods are devised. Here we showed how from time series we can extract the coordinates of points in phase space, and through the filtration of Čech complex the persistent homological structure can be recovered.

Bibliography

[1] MALETIĆ S, RAJKOVIĆ M, VASILJEVIĆ D. Simplicial complexes of networks and their statistical properties [J]. Lecture Notes in Computational Science, 2008, 5102(II): 568–575.

[2] MALETIĆ S, RAJKOVIĆ M. Combinatorial Laplacian and entropy of simplicial complexes associated with complex networks [J]. Eur. Phys. J. Special Topics, 2012, 212: 77.

[3] MALETIĆ S, HORAK D, RAJKOVIĆ M. Cooperation, conflict and higher-order structures of complex networks [J]. Advances in Complex Systems, 2012, 15: 1250055.

[4] BOCCALETTI S, LATORA V, MORENO Y, et al. Complex Networks: Structure and Dynamics [J]. Phys. Rep., 2006, 424: 175.

[5] NEWMAN M E J. Networks: An introduction [M]. Oxford: Oxford University Press, 2010.

[6] EARL C F, JOHNSON J H. Graph theory and Q-analysis [J]. Environment and Planning B, 1981, 8: 367.

[7] KOZLOV D. Combinatorial algebraic topology [M]. Heidelberg: Algorithms and Computation in Mathematics, Springer-Verlag, 2008.

[8] ANDELKOVIĆ M, TADIĆ B, MALETIĆ S, et al. Hierarchical sequencing of online social graphs [J]. Physica A., 2015, 436: 582.

[9] HORAK D, MALETIĆ S, RAJKOVIĆ M. Persistent homology of complex networks [J]. J. of Stat. Mech., 2009, 03: P03034.

[10] LOVÁSZ L. Kneser's Conjecture, chromatic numbers and homotopy [J]. J. Comb. Th. A, 1978, 25: 319.

[11] ARENAS F G, PUERTAS M L. The neighborhood complex of an infinite graph [J]. Divulgaciones Matematicas, 2000, 8: 69.

[12] DONG X, WACHS M L. Combinatorial Laplacian of the matching complex [J]. Electronic Journal of Combinatorics, 2002, 9: R17.

[13] MUNKRES J R. Elements of algebraic topology [M]. California: Addison-Wesley Publishing, 1984.

[14] HATCHER A. Algebraic topology [M]. Cambridge: Cambridge University Press, 2002.

[15] EDELSBRUNNER H, HARER J L. Computational topology: An introduction [M]. Providence: American Mathematical Society, 2010.

[16] DESILVA V, GHRIST R. Coverage in sensor networks via persistent homology [J]. Algebraic & Geometric Topology, 2007, 7: 339–358.

[17] DESILVA V, CARLSSON G. Topological estimation using witness complexes [C]. In Symp. on Point-Based Graphics, 2004: 157.

[18] GARLAND J, BRADLEY E, MEISS J D. Exploring the topology of dynamical reconstructions [J]. arXiv:1506.01128v1.

[19] MALETIĆ S, ZHAO Y, RAJKOVIĆ M. Persistent topological features of dynamical systems [J]. Chaos., 2016, 26: 053105.

[20] KANTZ H, SCHREIBER T. Nonlinear time series analysis [M]. Cambridge: Cambridge University Press, 1997.

[21] TAKENS F. Detecting strange attractors in turbulence [M]//RAND D, YOUNG L S. Dynamical systems and turbulence, Warwick 1980. Berlin, Heidelberg: Springer, 1981.

[22] SAUER T, YORKE M, CASDAGLI M. Embedology [J]. Journal of Statistical Physics, 1991, 65: 3.

[23] FRASER A M, SWINNEY H L. Independent coordinates for strange attractors from mutual information [J]. Phys. Rev. A, 1986, 33: 1134.

[24] GRASSBERGER P, PROCACCIA I. Measuring the strangeness of strange attractors [J]. Physica D, 1983, 9: 189–208.

[25] CARLSSON G. Topology and data [J]. American Mathematical Society Bulletin, New Series, 2009, 46: 255.

4

Several Applications of Simplicial Complexes

Previous chapters were devoted to preparing the stage for more thorough examples which will illustrate the potential for practical applications of simplicial complexes, and the examples do not exhaust all possible applications. On the other hand, from the contents of previous chapters, as well as from simple examples, the reader can already anticipate the possible applications.

Following the form of our exposition through last two chapters, it is noticeable that with each following chapter and section we introduce more and more detailed and complicated concepts, which finally result in the applications in concrete situations. We have defined simplicial complexes in three equivalent ways, namely we defined geometrical, abstract and via relation simplicial complexes. Clearly, depending on the particular problem which we want to tackle and on the specificities of the system under study, we will choose a convenient definition of simplicial complex. On the other hand, the way of building simplicial complex does not affect the choice of tools and methods for characterizing it, and hence we are provided with quite general framework for comparison between different complex systems. What often matters for the application of specific algebraic topology tools is whether we will take into consideration all the high-dimensional simplices, like in Q-analysis or combinatorial Laplacian, or they are to some extent only computational burden, like in homology group computation.

But knowing how to define mathematically simplicial complexes, and how to characterize them, is not enough. In order to implement application of simplicial complexes and algebraic topology, we need the ways to relate the phenomenon (that is the specific complex system) to the particular type of simplicial complex. Hence, we introduced different ways to build simplicial complexes from data, either already structured to some degree, like complex networks, or distributed in some space and we want to reconstruct the shape of data using, say, the Čech complex or the Vietoris-Rips complex.

First, as an example of the convenience of simplicial complex defined via relation to capture the social environment, we will introduce an opinion exchange model. The advantage of such approach is twofold: representing simplices as opinions is close to reality, and it is easy adjustable to the real world data, like the data gathered in public surveys. Next, we will build simplicial complexes from real world complex networks, and emphasis the emergence of mesoscopic structures, that is substructures built by the groups of nodes, within complex networks which cannot be restored using graph-theoretic tools only. Furthermore, we will demonstrate the suitability of different simplicial complexes built from the same network. Next, we show the usefulness of persistent homology in the analysis of time series obtained from the evolution of dynamical complex system, which may display chaotic behavior. And finally, based on everything that is introduced we will give an overview of the potential applications of simplicial complexes and algebraic topology to the aeronautical science in general.

4.1 Opinion exchange modeling

The content of this section relies mostly on the research ideas introduced in papers[1-3] whereas the purpose of this section is to exemplify the usage of simplicial complexes defined by relation and their suitability to capture the real world situations related to social interactions. Furthermore, the simplicial approach in modeling social environment is easy adjustable to the data gathered in public surveys, and may serve as computational framework for forecasting the social behavior.

Introduction

Let us start with an example which will help us to set the stage for introduction of an opinion exchange model. Imagine two colleagues who work in the same company, meet in the restaurant during lunch break at work, and start talking about personal well-being. Of course, each of them has his own opinion what is important for a person to be happy and satisfied with his life, and the reasons for holding a particular opinion can be various. For example, if one of them thinks that *physical health, personal finance,* and *marriage relation* are essential for personal well-being, whereas the other thinks that *social relations, shopping, work & career,* and *going on holidays* are important, then their opinions obviously does not have anything in common. Hence, it may be hard for them to achieve an agreement or consensus about some common opinion. On the other hand, suppose that the first colleague thinks that *physical health, personal finance,* and *work & career* are important for personal well-being, and the other colleague thinks that *personal finance, work & career, social relations,* and *going on holidays* are important, then their opinions overlap (sharing properties *personal finance,* and *work & career*), and they are more inclined to exchange their opinions. After all that is written about simplicial complexes, with this example we already have a hint about the way how to capture an opinion and its features by a simplex. But we will come back to this point soon.

The necessity to understand the social dynamics from opinion exchange to language formation resulted in building a variety of models.[4] Most often opinion models focus on formation of social dynamics of agents associated with certain property, which is changed under some dynamical mechanism controlled by one or more parameters. Like in our opening example, the opinions are characterized by certain number of properties and people sharing these properties have similar opinions. A simple model that is proposed here is used to illustrate the convenience of simplicial (and accordingly algebraic topological) approach to opinion dynamics, as well as to promote a topological setting grounded on the structure of a simplicial complex as the natural framework for opinion formation.

The interest of interdisciplinary research community in tackling the problems related to social dynamics delivered many simple agent-based models.[4] The most popular are, for example, the Voter model,[5] the Galam model,[6] the social impact model,[7] the Sznajd model,[8] the Deffuant model,[9] and the Krause-Hegselman model.[10] Some of these models consider population of individuals (in agent-based modeling called the *agents*) with discrete opinions represented as integers, for example $(+1/-1)$,[5–8] while others consider population of individuals with continuous, bounded range of opinions.[9, 10] In the so called consensus models[4] computer simulation of opinion dynamics starts with randomly distributed opinions over the population of agents located on the nodes of a graph. Interaction between agents differs from model to model and at the end of simulation the state of the agents may be characterized as consensus (single opinion state), polarization (two opinions), or anarchy (diversity of opinions). Researchers develop a model based on some social property, which originate from the socio-psychological research, characterized by interactions between agents which may include social phenomena characterizing communication situations such as social influence, homophily, transmission of information from the individual to his (her) neighbors, and bounded confidence.[4] In the current example of simplicial modeling of opinion dynamics we will restrict only on the bounded confidence criteria. The final benefit from this research can be the insight into the public environments related to social issues, which gives to the practitioners of the various kinds a common environment for taking social actions.

An individual may have an opinion about different issues, such as treatment of criminals or expansion of crime, religion, political issues including political parties, globalization, animal rights, global warming, advertising, movies, ... or nearly anything. An individual's social actions reflect his (her) opinions as sets of beliefs about certain issues (or generally subjects in their broadest meaning) which may or may not have an empirical background. In this sense the opinions are inferred from an individual's perception and knowledge about the likelihood of events or relationships regarding a specific topic. They also may involve evaluations of an event or an object.[11] In order to formulate or express an opinion on certain topic an individual evaluates a set of intermingled features or traits. When, for example, a person talks, writes an email or text of any kind about his (her) opinion, he (she) actually sends a "package" of interconnected traits which forman opinion. In this sense an opinion is not just a collection of independent traits but interrelated traits which form an opinion as a *whole*, in the sense that the meaning of specific opinion transcends the meaning of each judgment individually. If we go back to the example of conversation between two colleagues from the beginning of this section, we notice that the opinion of each of the colleagues is an authentic entity, and not just the mesh of different traits. When the conversation topic is the personal well-being a person expresses his opinion using all traits in an interconnected manner and juxtaposes them under a certain relationship. Adding a new trait to the existing opinion about person's well-being, a person's opinion shifts to the new opinion. A collection of opinions of a large number of individuals represents a mixture of overlapping opinions and shared traits, the analysis of which requires a suitable mathematical framework which captures the essence of opinions and their formation.

We consider an opinion as a simplex and the opinion set as a simplicial complex through shared traits build the overlapping connectivity structure of opinions. The simplicial complex of opinions also encloses an opinion space defined by the complex itself. The number of traits that characterizes a single opinion does not have to be the same and it may vary from opinion to opinion. Back to our example of person's well-being discussion. The first case is when two persons have different opinions (Fig. 4.1(a)), and it can be represented as two nonoverlapping geometrical simplices (Fig. 4.1(b)), whereas the case when two persons have different but overlapping opinions (Fig. 4.2(a)) can be represented as two overlapping simplices (Fig. 4.2(b)).

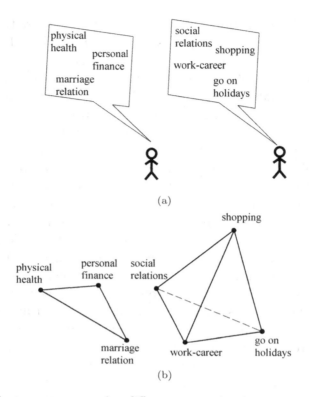

Fig. 4.1   Two persons expressing different nonoverlapping opinions about the personal well-being (a), and the associated simplicial representation (b)

Another convenience of representing opinions as simplices is in an easy implementation of the bounded confidence criteria[4] which, in the broad context, means that in order to interact, two individuals must not be too different. In the model which we will consider in this book, we will implement only the bounded confidence criteria for agents to communicate (i.e., to exchange opinions), and as we will see, even this simple criteria may lead to the interesting results.

Model

To recap, social entities in the form of individuals (or agents) are the carriers of social interpersonal interaction,[12] hence an opinion of a

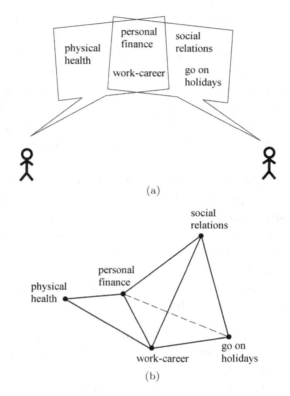

Fig. 4.2 Two persons expressing different overlapping opinions about the personal well-being (a), and the associated simplicial representation (b)

single person or a group of persons may be under the influence of another individual or another group of individuals. Interactions may be of dyadic type (between two persons), triadic type (between three persons) or in general of n-adic type which includes n persons and clearly social interaction depends on the interpersonal connectivity of persons. In the current model, we restrict ourselves only on dyadic interactions. The result of the interaction largely depends on whether two agents have the dissimilar or similar opinions, that is, if they have different but to a certain extent overlapping opinions.

An aggregation of opinions is associated to the aggregation of overlapping simplices, that is simplicial complex, and accordingly, the similarity between opinions is characterized by the degree

of overlapping between two simplices. In the current model we
assume that every person can communicate with every other person. Although each agent can interact with every other agent, an
agent communicates with another agent only if the bounded confidence criteria is satisfied, that is we take into consideration that two
agents can have a successful interaction only if their opinions do not
differ significantly. Accordingly, the simplicial representation of opinions has been found suitable for capturing the bounded confidence
criteria.

Let us associate the set T to traits and the set Ω to opinions, the
relation λ assigns opinions to the traits which form it, and then opinions build a simplicial complex of opinions.[2,3] We take the relation
λ as random, meaning that to an opinion ω_i is assigned randomly
with probability p a trait t_j. In this way we build a random Atkinian
simplicial complex.

Let n and m be the numbers of opinions and traits, respectively.
The simulation steps are as follows:

(1) generate a simplicial complex for particular values of n, m, and p;
(2) associate to each agent α an opinion ω_i^α (which is an integer
    between 1 and n);
(3) randomly choose an agent α with opinion ω_i^α and an agent β
    with opinion ω_j^β, then

    (a) if agents α and β have the same opinion nothing happens,
        and we choose another pair of agents;
    (b) if agents have different but overlapping opinions (that is,
        simplices associated to two opinions share a face), and the
        overlapping degrees of opinions, defined in[2] ($\theta_i = \frac{f_{ij}}{q_i}$ and
        $\theta_j = \frac{f_{ij}}{q_j}$, where f_{ij} is the dimension of sharing face, and
        q_i and q_j are the dimensions of opinions ω_i^α and ω_j^β, respec-
        tively) are less than the bounded confidence threshold ε, i.e.,

        $$\max(\theta_i, \theta_j) \leq \varepsilon$$

        then randomly the agent α will accept the opinion of agent
        β (that is ω_j^β) or the agent β will accept the opinion of agent
        α (that is ω_i^α).

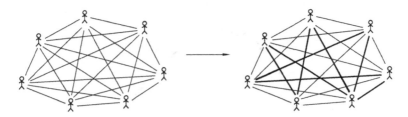

Fig. 4.3 Building complex network at the end of simulation. Links (in thicker lines) between agents which communicated more frequently and have the majority opinion are kept

(4) At the end of the simulation we calculate the degree distribution of network which remains after deleting all agents who don't hold the majority opinion, and links between agents who did not communicate frequently (Fig. 4.3). It is expected that different values of the bounded confidence threshold ε and simplex-vertex relation probability p will affect the structure of the final network, and accordingly its degree distribution.

Repeat steps (3) and (4).

Results

For considering the public opinion formation, and the reconstruction of underlying social network, we have considered different cases characterized by different values of simplex-vertex probability p. Namely, in order to make a successful comparison, and to draw meaningful conclusions, the number of opinions ($n = 6$), traits ($m = 60$), simplex-vertex relation probabilities ($p = 0.3, 0.5, 0.7$, which are chosen appropriately with respect to the results of dependence of simplicial structural complexity on relation probability, presented in the following), and the number of agents ($N = 1\,000$) does not change in each case, whereas the only varying parameter is the bounded confidence level (ε). We assume that the underlying network of agents is unknown a priori, and hence every agent may communicate with any other agent. All results are averaged over 100 simulations, and take equal iteration number (10^5).

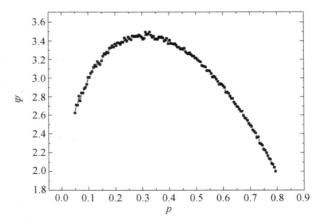

Fig. 4.4   The changes of simplicial structural complexity Ψ with the change of probability of relation p, for the vertex set size $m = 60$ and the number of simplices $n = 6$

It turns out that the simplicial structural complexity Ψ for simplicial complexes is a concave function of probability of relation p[3], as it is illustrated in Fig. 4.4 for simplicial complexes built from $n = 6$ simplices and $m = 60$ vertices. From the Fig. 4.4 it is obvious that our choice of simplex-vertex probabilities $p_1 = 0.3$, $p_2 = 0.5$, and $p_3 = 0.7$ satisfies $\Psi_{p_1} > \Psi_{p_2} > \Psi_{p_3}$.

But how does simplex-vertex probability p affect the structure of simplicial complex? Recall that p is the probability that the vertex belongs to the simplex. Hence, increasing the probability p, for fixed numbers of simplices and vertices, the dimensions of simplices increase, as well as the dimensions of shared faces between simplices. In the context of simplicial complex of opinions, that means that the opinions are more overlapping when increasing p, and hence it affects the bounded confidence criteria for opinion exchange. To illustrate this relation, consider the dependence of fraction of agent which hold the majority opinion F_{m} on the bounded confidence level as it presented in Fig. 4.5. For different values of probability ($p = 0.3, 0.5, 0.7$) we can notice that F_{m} is an increasing function of ε, and that there exists a shift to right on the ε-axes when we increase probability p. The latter result indicates that there is strong coupling between

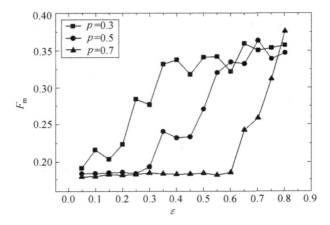

Fig. 4.5 The change of fraction of agents which hold the majority opinion F_m by changing bounded confidence threshold ε for three simplicial probabilities $p_1 = 0.3$ (squares), $p_2 = 0.5$ (circles), and $p_3 = 0.7$ (triangles)

probability p and the bounded confidence level ε, and hence between the simplicial structural complexity Ψ and the bounded confidence level ε.

From Fig. 4.5 we can notice that the fraction of agents who hold the majority opinion is 40%, and implicates that other opinions are present in society, hence the complete consensus is not achieved. Nevertheless, we can build a complex network under the criteria described in the section *Model*, and get an insight into the structure of the core of the fraction of agents who hold the majority opinion. From Fig. 4.6 we can see that the degree distributions $P(k)$ of networks for different values of simplex-vertex probability p follow the same shape, meaning that the structure of the obtained complex network independent of the structure of simplicial complex of opinions characterized (at least in our case) by simplex-vertex probability p.

Notes

Although our intention here is not to consider any particular real world situation, but rather to illustrate the way simplicial complexes

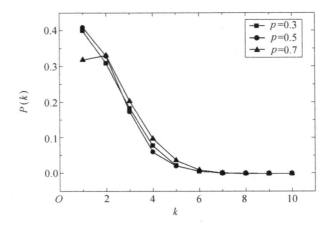

Fig. 4.6   Degree distributions of reconstructed complex social networks at the end of simulation for three simplicial probabilities $p_1 = 0.3$ (squares), $p_2 = 0.5$ (circles), and $p_3 = 0.7$ (triangles)

fit in the research of opinion dynamics, we obtained some interesting results. An opinion dynamics model which assumes that opinions are unordered sets of traits is developed. Namely, representation of opinions as multidimensional simplices is close to our intuition that creation of an opinion is the result of integration of traits which we deduce either through perception or learning, and hence the overlapping between two opinions is captured by sharing the same traits. In this way, the simplicial complex of opinions through the overlapping between opinions represents the potential communicative level for social interaction captured by the bounded confidence criteria. Increasing the bounded confidence level, the likelihood for opinion exchange between two persons increases as well. We showed that the global outcome of social interaction under the bounded confidence criteria depends on the opinion-judgment probability. Further, we have calculated the degree distributions of complex networks formed by the members of society which hold the majority opinion and who communicated more frequently between themselves, which resulted in the same shape of degree distributions.

In this simple model we treated the social dynamics as self-organized opinion exchange, in the sense of absence of external influences, and hence as an upgrade of the model we can introduce the mass media, by representing it as an external simplicial complex. Further, the simplicial complex representation of opinions can be easily related to the public opinion surveys, where questionnaires can be formulated in a way to adapt to the two-set definition of simplicial complex.

An opinion dynamics model which assumes opinions as unordered sets of judgments is developed. The opinion space is mapped to the simplicial complex so that the analysis uses concepts of combinatorial algebraic topology.

4.2 Topological description of the air traffic control network

In the previous section we illustrated how simplicial complexes can serve as a model of opinion space in a way that they are built from the relation between two sets. Now we will move further, and focus on a system which already has some structure, that is, a complex network. In that course, we will start with real world complex network and build three simplicial complexes (the neighborhood complex, the clique complex, and the conjugate clique complex) in a way it is introduced in the Chapter 3, and then we will calculate some of the quantities defined in the Chapter 2. As we have indicated in previous chapters, we will show that through building different simplicial complexes from a single network different mesoscopic structures emerge giving us more possibilities to get an insight into the structure of underlying network.

The content of this section is mainly based on the work presented in research papers,[13–16] although in some research papers the quantities are calculated for different purpose. Nevertheless, the general conclusions about the results of this section can be easily related to the conclusions drawn in referenced papers. Our intention here is not to tackle any specific problem, but to present the similarities and the differences between different substructures which emerge in complex networks. Hence, in some sense we set a scene for further research which would serve as an upgrade of the results presented.

Elementary network properties

We will build simplicial complexes from the Air Traffic Control network (Fig. 4.7) in which nodes represent airports or service centers, and links represent preferred routes recommended by the the USA's FAA (Federal Aviation Administration) National Flight Data Center (NFDC), Preferred Routes Database. The data source of the network can be found on the Internet, whereas elementary data about this network can be found on the Internet. For our current consideration it is enough to know that the number of nodes is $N = 1\,226$, and the number of links is $L = 2\,615$. Although the original network is directed, we will consider only undirected case. The degree distribution of links displays power-law behavior with an exponent -2.35 (drawn on log-log scale in Fig. 4.8), and hence we can say that it belongs to the class of scale-free networks.[17] This fact is particularly interesting since the scale-free networks are of special interest in the complex networks' research (see for example Ref. [18]). The more detailed analysis of properties of this network, which emerge from scale-freeness, are out of the scope of this book, and hence, we

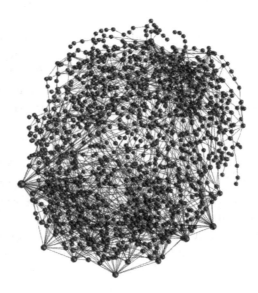

Fig. 4.7 Graphical representation of air traffic control network

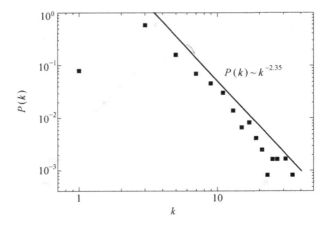

Fig. 4.8   Degree distribution of air traffic control network. The power-law fit on log-log scale with an exponent −2.35 is indicated

will restrict ourselves on the fact that degree distribution displays power-law.

Neighborhood complex

We have built the neighborhood complex[19, 20] from the Air Traffic Control network using the recipe introduced in the Chapter 3. Recall that in neighborhood complex vertices are nodes from the original graph, and to each node i from the original graph is associated a simplex σ_i in the neighborhood complex, defined by vertices which are linked (i.e., adjacent, neighbors) to i in the original graph. For the purpose of introduction of basic simplicial properties, we have calculated the Q-vector, the Second structure vector, the eccentricity, and the vertex significance. In Fig. 4.9 we have illustrated on log-log scale the changes of Q-vector and the Second structure vector[21] values with the change of q-levels. Even from a simple visual inspection we can notice that both quantities display some sort of regularity, especially between 2-level and 30-level where it is best approximated by the power-law fit. The value of Q-vector at 0-level indicates that the structure of the neighborhood complex, and therefore, of the original complex network is connected, that is $Q_0 = 1$.

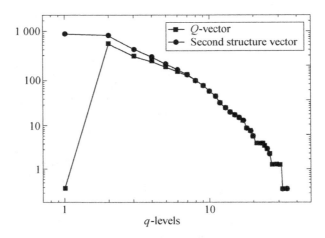

Fig. 4.9   The values of the Q-vector (i.e., the first structure vector) and the second structure vector of neighborhood complex built from the air traffic control network

From the distribution of eccentricity[22] $P(ecc)$ values of simplices (Fig. 4.10(a)) we can notice that one peak dominate, that is the one around eccentricity value $ecc = 0.4$ indicating that $\approx 45\%$ of simplices have approximately the following relationship between bottom $q(\hat{q})$ and top $q(\hat{q})\check{q} = \frac{(3\hat{q}-2)}{5}$, and recall from the Chapter 2 that $\check{q}$ is the largest dimension of faces which simplex share with other simplices. On the other hand, the peak around the eccentricity value $ecc = 0$ indicate that around 13% of simplices are completely integrated into the structure, that is they are the faces of other simplices. The distribution of vertex significance[22] $P(vs)$ values illustrated in log-log plot in Fig. 4.10(b) displays approximately power-law behavior, which means that vs inherits the behavior of degree distribution. This finding is in accordance with the results from.[13, 14]

Clique complex

Using Bron-Kerbosch algorithm[23] we find all maximal cliques in the Air Traffic Control network, and build the clique complex.[24] The results of calculation of the Q-vector and the Second structure

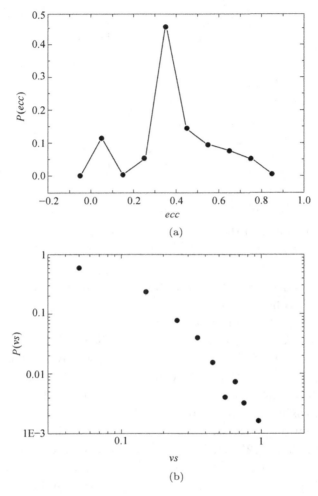

Fig. 4.10 Distributions of the eccentricity (a) and the vertex significance (b) of neighborhood complex built from the air traffic control network

vector[21] values, as it is illustrated in Fig. 4.11, show that the dimension of simplicial complex is 4, i.e., the largest clique in network is formed by 5 nodes. The overlapping of values for both vectors at 2-, 3- and 4-levels indicate that at these q-levels the structure is disconnected, that is, at these levels each connectivity class is formed by only one simplex, and there is not any pair of simplices which share

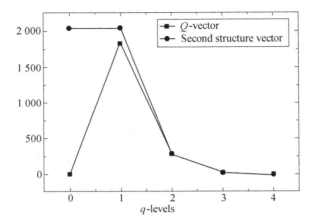

Fig. 4.11   The values of the Q-vector (i.e., the First structure vector) and the second structure vector of clique complex built from the air traffic control network

3 or more vertices. Since the number of q-levels is rather small, it is meaningless to fit these data.

The distribution of eccentricity[22] $P(ecc)$ illustrated in Fig. 4.12(a) displays an interesting behavior. Namely, two dominating peaks emerge: the first for the eccentricity value $ecc = 0.33$, and the second for $ecc \approx 0.5$. Particulary interesting is that almost 90% of simplices have the value $ecc \approx 0.5$, which means that their maximal face is approximately two times less than the dimension of simplex. The distribution of vertex significance[22] $P(vs)$ illustrated on semilog scale in Fig. 4.12(b) displays exponential behavior. If we go one step back, and recall the way we built clique complex, that is, by finding all maximal cliques, the value of vertex significance actually tells us how much a clique is important with respect to the vertices which form it. Hence, the distribution of importance of cliques with respect to the nodes which build them is following the exponential behavior.

Conjugate clique complex

In the previous section we analyzed simplicial complex build from maximal cliques of the Air Traffic Control network. Now, reversing

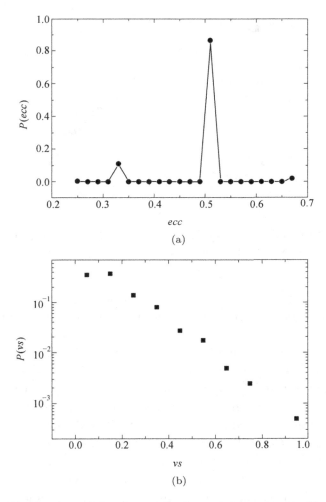

Fig. 4.12 Distributions of the eccentricity (a) and the vertex significance (b) of clique complex built from the air traffic control network

the roles of cliques and nodes by assigning simplices to the nodes, and vertices to the cliques, we build the conjugate clique complex of the Air Traffic Control network, following the method from the Chapter 3. Although the Q-vector and the Second structure vector[21] display approximately power-law behavior, from Fig. 4.13 we can notice that for 25 q-levels (from 28- to 4-level) all connectivity classes

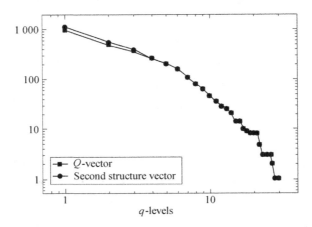

Fig. 4.13 The values of the Q-vector (i.e., the first structure vector) and the second structure vector of conjugate clique complex built from the air traffic control network

contain only one simplex. In other words, going again one step back, it means that nodes can have the common membership to four cliques or less.

From the distribution of eccentricity[22] $P(ecc)$ displayed in Fig. 4.14(a) we can notice that more than 45% of simplices have the value of eccentricity approximately $ecc = 0.45$, whereas the fraction of simplices which have the values $ecc \approx 0$, $ecc \approx 0.65$, and $ecc \approx 0.75$ together is approximately 37%. The distribution of vertex significance[22] $P(vs)$ illustrated on log-log scale in Fig. 4.14(b) displays approximately power-law behavior, especially for the values of vertex significance between $vs = 0.1$ and $vs = 1$. Going one step back, the value of vertex significance quantifies the importance of a node in the original network with respect to the cliques to which belongs, and hence, the distribution of importance of nodes with respect to the cliques to which it belongs is following the power-law behavior.

Combinatorial Laplacian

We have calculated the eigenvalue spectra of combinatorial Laplacian[25] of clique complex only, although it can be calculated for the conjugate clique complex and the neighborhood complex as well,

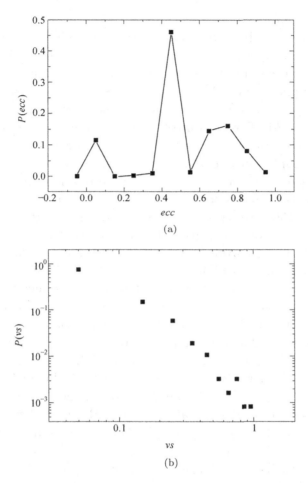

Fig. 4.14 Distributions of the eccentricity (a) and the vertex significance (b) of conjugate clique complex built from the air traffic control network

which can give additional interesting results. For the graphical illustration of eigenvalue spectra we have used the Cauchy-Lorentz kernel for the convolution of the spectral density yielding the following density function[26]

$$f(x) = \sum_i \frac{\gamma}{(\lambda_q^i - x)^2 + \gamma^2}$$

where we have chosen the parameter γ to have the value $\gamma = 0.03$.

As we mentioned in the Chapter 2, the eigenvalue spectra of 0-dimensional combinatorial Laplacian is actually the graph Laplacian. Hence, it would be constructive to give some comparison, as well as to establish potential relationship between the eigenvalue spectra of graph Laplacian and the eigenvalue spectra of higher-order combinatorial Laplacian. Since we are building the simplicial complex from cliques of the Air Traffic Control network, which aggregate into network's substructures, than establishing some relationship between the graph Laplacian and the higher-order combinatorial Laplacian actually means that we are establishing the relationship between the graph and the mesoscopic structures embedded in it.

In Fig. 4.15(a) we represented the spectral plot of 0-th combinatorial Laplacian, that is, the graph Laplacian. We will pay a particular attention to the eigenvalues $\lambda_1 = 0$, $\lambda_2 = 2$, $\lambda_3 = 3$. Although it is not obvious from this Figure, but the multiplicity of $\lambda_1 = 0$ eigenvalue is 1, which means that the structure is connected, i.e., has only one component. This result is in accordance with the results obtained from calculation of Q-vectors for all considered cases (the neighborhood complex, the clique complex, and the conjugate clique complex), and it also means that the 0-th Betti number is equal to $\beta_0 = 1$. When we look now at the spectral plot in Fig. 4.15(b) for 1-st combinatorial Laplacian, first we can notice that there are many $\lambda_1 = 0$ eigenvalues, that is the 1-st Betti number is equal to $\beta_1 = 875$, meaning that there are 875 1-dimensional holes in the network. Second, we can notice that the eigenvalues $\lambda_2 = 2$, $\lambda_3 = 3$, and $\lambda_4 = 4$ appear as well. Hence, the eigenvalues $\lambda_2 = 2$, and $\lambda_3 = 3$, appear in both spectra (graph Laplacian and higher-order Laplacian), and considering only graph Laplacian is not enough to track the origin of these two eigenvalues. And if we proceed further to the spectra of 2-nd combinatorial Laplacian, the characteristic peaks in Fig. 4.16(a) appear for eigenvalues $\lambda_2 = 2$, $\lambda_3 = 3$, $\lambda_4 = 4$ and $\lambda_5 = 5$ indicating that the origin of the eigenvalues $\lambda_2 = 2$, and $\lambda_3 = 3$ in the graph Laplacian spectrum (Fig. 4.15(a)) may be related to the higher-order structures in complex network captured by the higher-order combinatorial Laplacian. For the eigenvalue $\lambda_3 = 3$ it is particularly noticeable as it appears in combinatorial Laplacians from 3-rd

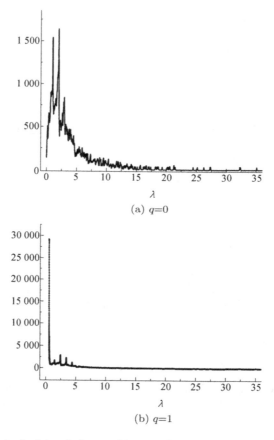

Fig. 4.15   Spectral plots of the combinatorial Laplacian's eigenvalues of clique complex built from the air traffic control network for dimensions for $q = 0$ (a) and $q = 1$ (b)

(see Fig. 4.16(b)) to 0-th. A similar feature can be noticed for some eigenvalues which does not appear in the graph Laplacian eigenvalues, but appear in the higher-order Laplacian spectra. Namely, the appearance of the eigenvalue $\lambda_5 = 5$ is noticeable in the spectra of 4-th (Fig 4.16(c)), 3-rd (Fig. 4.16(b)), and 2-nd (Fig. 4.16(a)) combinatorial Laplacian. The origin of $\lambda_5 = 5$ eigenvalue comes from the existence of 4-dimensional simplex which is the reason that for 4-th combinatorial Laplacian (Fig. 4.16(c)) we have only one eigenvalue.

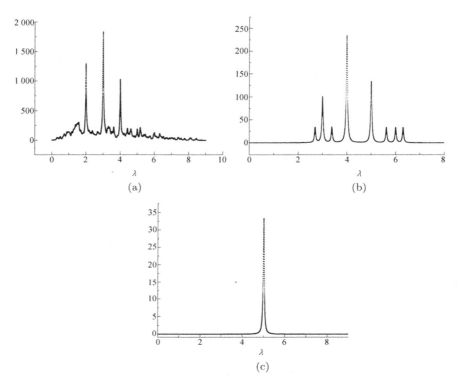

Fig. 4.16   Spectral plots of the combinatorial Laplacian's eigenvalues of clique complex built from the air traffic control network for dimensions for $q = 2$ (a), $q = 3$ (b), and $q = 4$ (c).

Recall that the eigenvalue of a single q-simplex is equal to the number of vertices which build that simplex, that is $q + 1$, and if we compare this result with the examples of spectra introduced in the Chapter 2 illustrated in Fig. 2.22, it is easy to deduce that the persistence of the eigenvalue $\lambda_5 = 5$ originates from the 5-clique to which the 4-simplex is associated, and which overlaps with other simplices only at the lower levels of connectivity.

We can notice from the Figs. 4.15 and 4.16 that many different eigenvalues appear in the spectra of all combinatorial Laplacians (except in the spectra of 4-th), whereas we did only the analysis of few characteristic eigenvalues. Since our attention is to scratch the versatility of applications, any further deeper analysis of combinatorial Laplacian spectra is out of the scope of book.

Notes

This section is overwhelmed with various results, all related to the various properties of a single real world complex network: the Air Traffic Control network. As we have already given a hint in the Chapter 3 where we introduced different ways of building the simplicial complex from a graph, in this section we illustrated it in practice through building the neighborhood complex, the clique complex, and the conjugate clique complex.

Although the analysis presented in this section is not comprehensive as it can be, there are some interesting findings which we will highlight. The behavior of the Q-vector and the Second structure vector, as well as the distribution of vertex significance for the neighborhood complex and the conjugate clique complex, display statistical invariance[13, 14] in the sense that their dependence on q-levels follows the behavior of the degree distribution of the underlying complex network. From the distribution of eccentricity the peaks emerge for characteristic *ecc* values and we related the maximal dimension of the shared face and the dimension of the associated simplex for each particular case.

We pointed out the importance of considering the higher-order combinatorial Laplacians, as well as demonstrated the relationship between eigenvalues which appear in the spectra of different combinatorial Laplacians.[15, 16] Furthermore, the emergence of some eigenvalues in graph Laplacian spectra has its origin in the relationships between high-order aggregations of nodes in a complex network, and therefore, the analysis of only graph Laplacian is not enough for understanding the structural properties of the underlying complex network.

4.3 Topological properties of dynamical systems

In previous two sections we presented the examples of applications where it is assumed that the data set is given, and then simplicial complex is built directly from it. Namely, in the opinion model simplices and vertices are explicitly related to the opinions and traits, respectively, whereas in building simplicial complexes from complex network vertices are nodes of the network, and then we apply different criteria to build simplices as aggregations of nodes of the underlying network. Now we move forward to the case where simplicial complex is not built directly from the data set, but we need to introduce an intermediary step.

In the Chapter 3 we sketched a way to build the simplicial complex from time series as a part of the method related to the phase space reconstruction of dynamical system. In this section we will apply that method which originates from the research presented in,[27] and show that the simplicial complex topology captures the topological structure of the underlying phase space. To accomplish that task, and make the method more understandable, we will build simplicial complex from the time series of the well-known dynamical systems.

The Ikeda map

As an example of 2-dimensional dynamical system which may display chaotical behavior we will use the Ikeda map[28, 29]:

$$
\begin{aligned}
x_{i+1} &= 1 + u(x_i \cos t_i - y_i \sin t_i) \\
y_{i+1} &= u(x_i \sin t_i + y_i \cos t_i)
\end{aligned}
\tag{4.1}
$$

where

$$
t_i = 0.4 - \frac{6}{1 + x_i^2 + y_i^2}
$$

and u is a parameter. The original equation can be found in a slightly different form, nevertheless the dynamical behavior of the system is the same. Although the parameter u can have different values, for a particular value of parameter $u \geq 0.6$ the system exhibits chaotic behavior with a characteristic shape of the strange attractor (Fig. 4.17). It is important to note that the Ikeda system

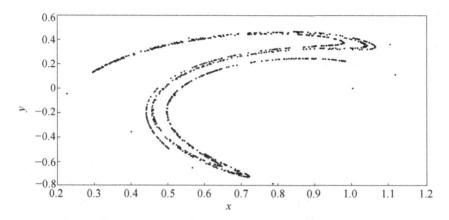

Fig. 4.17   The strange attractor of the Ikeda system

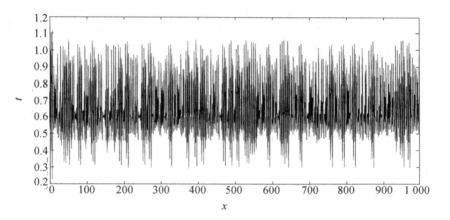

Fig. 4.18   Time series of the x-component of the Ikeda system

is 2-dimensional, and even from the visual inspection, it does not contain any holes in the topological sense.

We will apply the reconstruction method introduced in the Chapter 3 on the time series of the x-variable values was generated for the Ikeda system for the parameter value $u = 0.6$. The time series of the variable x is illustrated in Fig. 4.18.

The Rössler system

In the dynamical systems research one of the most popular bench-mark models is the Rössler system[30] represented by equations:

$$\dot{x} = -(y + z)$$
$$\dot{y} = x + ay \tag{4.2}$$
$$\dot{z} = b + xz - cz$$

where a, b and c are constants. When the values of parameters a and b are fixed (say $a = b = 0.2$) and the values of parameter c are tuned, the Rössler system displays significantly different behavior which is revealed in the emergence of a particular type of the attractor in the phase space, leading to the chaotic state characterized by the existence of the strange attractor for the value of parameter $c = 6.3$, which is presented in Fig. 4.19. For our subsequent considerations it is important to have in mind that the Rössler system is 3-dimensional system and that, even after visual inspection, the strange attractor is characterized by the emergence of a hole (see Fig. 4.19).

Like in the case if the Ikeda map, in order to apply the method introduced in the Chapter 3 we need to start with the time series of the single component and reconstruct the topological (or more precisely homological) structure of the Rössler system. In that course we have chosen the x-component for the application of our method, whose time series is illustrated in Fig. 4.20.

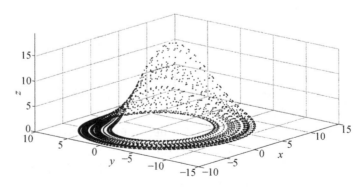

Fig. 4.19   The strange attractor of the Rössler system

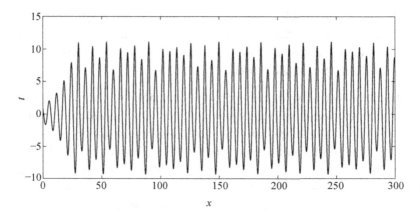

Fig. 4.20   Time series of the x-component of the Rössler system

Decimation of data set

From the simple inspection of Fig. 4.19 we can deduce that building simplicial complex from the points in phase space can result in enormous number of simplices. Furthermore, since we want to examine filtration of the Čech complex, by increasing the radius, and increasing the mutually overlapping balls, the high-dimensional simplices may emerge.[31] The emergence of both properties leads to the computational burden[32] which may hinder the application of method, and hence some sort of coarse-graining of data is necessary. In order to simplify and shorten calculation, and at the same time not to lose information about the phase space structure, we applied the following coarse-graining procedure introduced in[27]:

(1) Replace the group of 10 to 15 original data points in phase space by a point at the center of the smallest Euclidean ball which contains the original set of data points. The exact number of replaced data points depends on their distribution and density in the phase space.

(2) Then apply the Čech complex filtration on the new data set. Recall that the Čech complex is a simplicial complex which is formed such that a q-dimensional simplex is added when a subset of $q + 1$ points with common intersection of balls for certain

predetermined radius emerge. And also, as we emphasized in the Chapter 3 according to the Nerve theorem,[33] the Čech complex recovers the homotopical and homological features of the underlying topological space.

One way to test whether the new data points inherit the properties of the original data points is to calculate the so called correlation sum.[34] We will not go into detailed derivation of the correlation sum, but we will limit ourselves on the information that can be gained from calculating it. Namely, it is important for us to know that the correlation sum of the new and the original (that is, coarse-grained) data have the same scaling behavior for chaotic systems, due to the fractal nature of the strange attractors. For some set of N points in d-dimensional space $x_1, x_2, \ldots, x_N$ the correlation sum is defined by[34]

$$C_d(l) = \lim_{N \to \infty} \frac{1}{N^2} \sum_{i \neq j} \Theta(l - |x_i - x_j|)$$

where $\Theta(\cdot)$ is the Heaviside function. When the data size is infinite and l small, the correlation sum scales like $C(l) \sim l^D$, where D is called correlation dimension.

Persistent homology of phase space

Before applying the coarse-graining procedure and building the Čech complex from new data points, the time lag parameter (τ) is determined as the first minimum of the average mutual information.[35] Then m-dimensional vectors $(m = 2, 3, 4)$ are formed, and in that way we mapped the state vector of a dynamical system to a point in the reconstructed space. And finally, using the Čech complex filtration and the coarse-graining procedure we computed the persistent homology of Ikeda system and Rössler system.

The building of Čech complex can be understood in the following way. By increasing the radius around each data point the balls around points overlap for some radius values, and holes appear like the islands of inaccessability (i.e., obstacles). When the radius is increased further, holes either continue to exist or they disappear.

Persistent holes reveal the robust set of points which form a non-boundary cycles. The persistent holes display a signature of permanent obstacles, and the subspace of the phase space corresponding to the short-lived holes may be interpreted as temporary inaccessible. And further, increasing the radius we can separate topological noise (short-lived features) from the significant, long-lived and important topological features.

Application of the persistent homology method to the nonlinear dynamic systems requires additional parameter in the form of the embedding dimension. Thus, the homology generators (Betti numbers) are parameterized by the radius of the data cloud surrounding appropriate phase space points and by the embedding dimension which enables adequate reconstruction of the attractor. For small values of the radius r the complex is discrete set while for r fairly large it is a single high-dimensional complex. However, it would be misleading to try to determine the optimal value of the radius r, nor is it relevant to determine the number and type of holes appearing for each parameter r value. Hence, our interest is in the homology groups which contain significant holes which span through the largest interval of radius and whose persistence becomes pronounced at a certain, minimal embedding dimension. Consequently, if this persistence repeats at higher embedding dimensions, we have not only determined significant, long-lived topological features but also the minimal embedding dimension at which those features are manifested. Persistent diagrams make this distinction easier to reveal, and accordingly, we will perform our analysis using persistent diagrams, rather than using the persistent barcodes. Recall that in the persistence diagram ordinate is associated with radius values of the death of homology generators, whereas abscissa is associated with radius values of the birth of homology generators, and thus the coordinates of a point in persistence diagram represent birth and death of a generator (i.e., a q-dimensional hole) of the q-th homology group. The points near or on the diagonal in persistence diagram are associated with short-lived q-dimensional holes representing topological noise.[36]

We have noticed that in the strange attractor of Ikeda system does not contain holes, which lead to the expectancy that when we apply the Čech complex filtration we should not find any persistent holes. The confirmation is obvious from the inspection of the persistent diagrams for the 1-st homology group for 2-, 3-, and 4-dimensional embedding in Figs. 4.21–4.23, and hence the emergence of the topological noise (i.e., short-lived homology group generators) is expected. From Figs. 4.21–4.23 we can note that the same phenomena is repeated for different embedding dimensions. In the

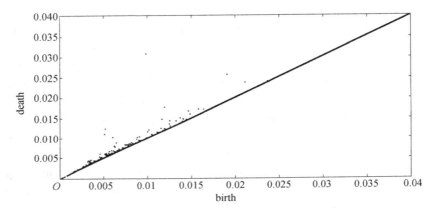

Fig. 4.21   Persistent diagram of the H_1 homology group of the Ikeda map for the 2-dimensional embedding dimension

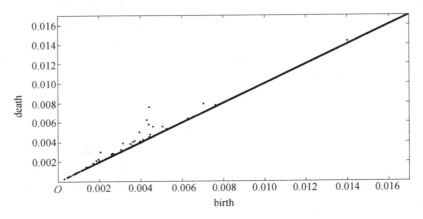

Fig. 4.22   Persistent diagram of the H_1 homology group of the Ikeda map for the 3-dimensional embedding dimension

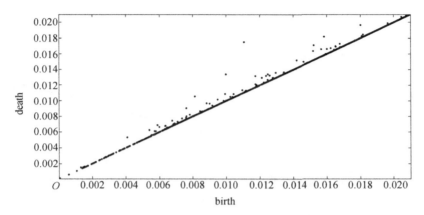

Fig. 4.23   Persistent diagram of the H_1 homology group of the Ikeda map for the 4-dimensional embedding dimension

case of the 2-dimensional embedding 1-holes persist longer than in the cases of 3-dimensional and 4-dimensional embedding, but not enough to escape the role of topological noise. The important conclusion is that, comparing Figs. 4.21–4.23 the phenomena of the topological noise occurrence and the lack of persistent homology group generators emerges even when the embedding dimension is increased, hence it follows this same qualitative pattern.

Let us check now whether the similar phenomena occurs in the case of the Rössler system as well. As we already noticed, a brief visual inspection of the attractor discloses one hole so it would be expected that the persistence diagram indicate their appearance. The persistent diagrams associated to the 2-, 3-, and 4-dimensional embeddings of the 1-st homology group generators are presented in Figs. 4.24–4.26, respectively. For the embedding dimension equal to 2 the persistence diagram for homology group generators of the 1-st homology group (Fig. 4.24) clearly indicate persistent lifetime of the one hole indicated by the circle. When the embedding dimension is increased to 3 (Fig. 4.25) and 4 (Fig. 4.26) their lifetime is preserved, that is they survive the filtration, as well as the increase of the embedding dimension. For practical reasons the death-value is chosen arbitrary although its value in terms of persistent homology is actually infinity. Hence, the topological structure is preserved.

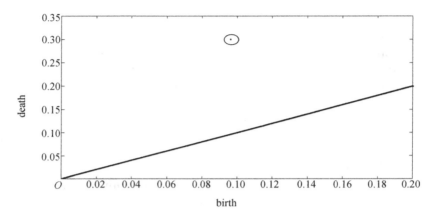

Fig. 4.24   Persistent diagram of the H_1 homology group of the Rössler system for the 2-dimensional embedding dimension; The persistent homology group generator is highlighted by circle

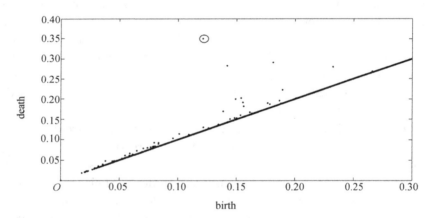

Fig. 4.25   Persistent diagram of the H_1 homology group of the Rössler system for the 3-dimensional embedding dimension; The persistent homology group generator is highlighted by circle

Notes

In this section we wanted to show that topological properties are preserved when we embed the data in the spaces of different dimensions. Two exemplar systems are chosen intentionally, since in one case we do not have persistent homology group generators (the Ikeda

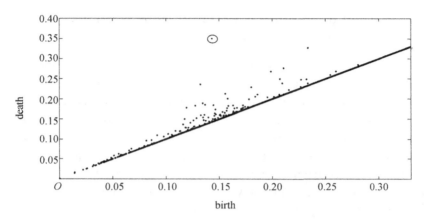

Fig. 4.26   Persistent diagram of the H_1 homology group of the Rössler system for the 4-dimensional embedding dimension; the persistent homology group generator is highlighted by circle

system), and in the other case (the Rössler system) the attractor is characterized by 1-dimensional hole, hence the persistence of 1-dimensional homology group generator is expected. Note that the basic features are observed even at low embedding dimensions, as it is observed in Refs. [32] and [27]. Actually, the reason is that preservation of topological features in the persistent homology reconstruction method requires only the property of homeomorphism in contrast to the diffeomorphism of the delay-reconstruction procedure. For example, the main macro topological features of the Rössler system, namely the one hole, is clearly distinguished for the embedding dimension as low as 2. In the Chapter 3 we did not mention the convenience, and sometimes the necessity, of the coarse-graining procedure, but building the simplicial complex after the application of the coarse-graining procedure makes computation more economical than some other similar techniques.

The reader may ask, and with a good reason, what kind of information can be retrieved from calculation of 0-th, 2-nd, and 3-rd persistent homology. Nevertheless, to avoid the confusion and the burden of large accumulation of results we didn't present the results for 0-th, 2-nd, and 3-rd persistent homology generators, although they display an interesting behavior too.[27]

4.4   Instead of summary: Potential contributions to aeronautical science

At the end of each chapter we gave a short summary of introduced concepts or methods. Nevertheless, since in this chapter we were focused on more concrete and more complex applications of simplicial complexes, instead of summary the last section has the role of convergence of different applications into a single research field, that is aeronautical science. For the purposes of current work we will accept the general definition of aeronautics as "a science that deals with airplanes and flying", hence we will consider different aspects like business, technology, management, aviation, ... related to aircraft.

As we have mentioned many times throughout the book, our introduction of theory and methods related to simplicial complexes, as well as examples of their applicability, does not exhaust all possibilities. And the same is the case with applications in aeronautical science, and we encourage the reader to find further applications. Namely, regarding the possible applicability in aeronautical science, we divided this section into two parts, depending on the applicability of the examples introduced in this section, or on the other applications of simplicial complexes.

First we have introduced the opinion exchange model and emphasized that it can be easily adjustable to the questionnaires of public surveys, due to the property that we can build simplicial complex from the relation between two sets. Now, suppose that big corporation that designs and manufactures airplanes wants to know the opinions of workers about the working conditions or whether they are satisfied with their well-being (refer to the example in opinion model section), which is largely dependent on the working conditions of employees. The results of such a survey may influence the change of company's policy so that the company can have more satisfied employees, but also the results can help company's Human Resource department in planning the recruitment programs since they will know how to attract new candidates. Hence, in very simplified version, they conduct a survey based on some questionnaire, and then the data are processed statistically. Finally, the interpretation of results influences making decisions and future plans of the company. On the other hand, something else can be done. If we go just one step back, i.e., represent the questions and answers as simplicial complexes, and since we are already equipped with the apparatus of, say, Q-analysis, we can get yet another set of useful information. Recall that the results of Q-analysis reveal the information about the structural relationships between data sets, and hence together with the statistical analysis, the analyst can get more detailed picture about the opinions of employees. Or the data can be used to simulate the exchange of opinions among employees. In either case the company can benefit from the outcome of analysis.

Next we were considering the Air Traffic Control network. Nevertheless, regarding the air traffic control, the application of simplicial complexes does not have to be restricted to building it from complex network. Basically, the task of air traffic controllers is to ensure safe operations of commercial and private aircrafts through, say, keeping them at safe distance from each other, and direct them during takeoff and landing from airports, and generally, to coordinate flights of thousands of airplanes. That means that there are many factors which influence the cognitive complexity,[37] meaning that if a controller becomes overloaded when he is experiencing the situation where the cognitive complexity exceeds his capabilities he can outperform leading to the threat to safety. Hence, it is of great importance to understand the mechanisms for reducing the cognitive complexity. Usually, focused interviews with air traffic controllers and traffic management unit personnel are conducted in order to understand those mechanisms, and we assume that the data from an interview can be adapted in such a way to build the simplicial complex, whose analysis (calculating the homology groups and/or Q-analysis) can reveal more insight in the cognitive framework of air traffic controller.

Back to the complex networks and the simplicial complexes. The Air Traffic Control network is not the only complex network that can be related to the air transportation, and there is one, even more important network, which has the influence on the society in many different ways. Namely, the Air Transportation network[38] (or the flight network) in which the nodes are the airports, which are connected by links when a direct flight is scheduled between them. We will illustrate the importance of Airline Transportation network on few examples. First, the robustness of air transport network[39] is of paramount importance, since the disruptions of communication links may cause serious problems for airlines, as well as for countries, or in other words, the whole society is affected. In order to reduce the effects of such disruptions it is important to understand the effects of possible causes of disruptions. Generally, the types of causes of disruption are considered: disruption of links by unintentional causes (like weather clemency) or disruption of links by intentional causes (like terrorist attacks on airports). Clearly, these disruptions affect the structure of the whole network, and accordingly they affect the relationships between higher-order structures (i.e., simplices) embedded in complex network. This lead us to the question: how does failures affect the topology of diverse simplicial complexes built from complex network? Recall that in the algebraic topological computations we are dealing mostly with different sorts of chains of simplices, and on the other hand simplices which are built from complex network represent groups of nodes, that is their aggregations build a sort of communities, called the simplicial communities.[15, 16] Hence, the research may be focused on the impact of disruptions on the topological properties of simplicial communities in complex networks, which can shed a new light on the resilience of networks, on the one hand, and can help the improvement of networks' robustness.

The robustness of the Air Transportation network is not the only property which is related to its structure. There is another important phenomenon which is of great importance for the whole society — that is the epidemics through the complex network.[40] Namely, due to the high frequency of mobility of passengers traveling large distances, the Air Transportation network enhances the global spread of diseases which may lead to pandemic. In order to control the spread of disease, the research is focused on the predictability of global epidemics caused by the structure of Air Transportation network and the flow of passengers.[41] Again, like in the case of network's robustness, by building different simplicial complexes, their topological analysis, and the analysis of flow of passengers over simplicial complexes, may contribute to better prevention of the epidemics, or if the spread of disease already started, to control its further development.

The third example which we considered is the application of apparatus of algebraic topology on the phase space reconstruction from time series of single variable. The generality of introduced method provides the applicability in any research in dynamical systems, which is widely present in the aeronautics research. For example, in the jet and rocket engine combustion research the time series of acoustic pressure can be used in analysis of the combustion dynamics,[42] and the characteristics of simplicial complex derived from time series may help researchers to better understand system's dynamics.

The developments of new and design of modern aircrafts are relying on the scientific computing. In other words, in order to understand the dynamics of complex flow patterns that is appearing in aviation-related problems, the simulations play the key role. Accordingly, the applicability of simplicial complexes as simulation techniques in Computational Fluid Dynamics (see, for example Ref. [43]) has an essential role in the aeronautics research.

Bibliography

[1] MALETIĆ S, RAJKOVIĆ M. Simplicial complex of opinions on scale-free networks [J]. Studies in Computational Intelligence, Springer, 2009, 207: 127–134.

[2] MALETIĆ S, RAJKOVIĆ M. Consensus formation on simplicial complex of opinions [J]. Physica A, 2014, 397: 111–120.

[3] MALETIĆ S, ZHAO Y. Hidden multidimensional social structure modeling applied to biased social perception [J], Physica A, 2018, 492: 1419–1430.

[4] CASTELLANO C, FORTUNATO S, LORETO V. Statistical physics of social dynamics [J]. Rev. Mod. Phys., 2009, 81: 591.

[5] CLIFFORD P, SUDBURY A. A model for spatial conflict [J]. Biometrika, 1973, 60: 581.

[6] GALAM S. Minority opinion spreading in random geometry [J]. Eur. Phys. J. B, 2002, 25: 403.

[7] NOWAK A, SZAMREJ J, LATANÉ B. From private attitude to public opinion: A dynamic theory of social impact [J]. Psychol. Rev., 1990, 97: 362.

[8] SZNAJD-WERON K, SZNAJD J. Opinion evolution in closed community [J]. Int. J. Mod. Phys. C, 2000, 11: 1157.

[9] DEFFUANT G, NEAU D, AMBLARD F, *et al.* Mixing beliefs among interacting agents [J]. Adv. Compl. Sys., 2000, 3: 87.

[10] HEGSELMAN R, KRAUSE U. Opinion dynamics and bounded confidence: Models, analysis, and simulation [J]. JASSS, 2000, 5(3): 498.

[11] OSKAMP S, SZHULTZ P W. Attitudes and opinions [M]. New Jersey: Lawrence Erlbaum Associates, Mahwah, 2005.

[12] KRECH D, CRUTCHFIELD R S. Individual in society: A textbook in social psychology [M]. USA: McGraw-Hill, 1962.

[13] MALETIĆ S, RAJKOVIĆ M, VASILJEVIĆ D. Simplicial complexes of networks and their statistical properties [J]. Lecture Notes in Computational Science, 2008, 5102(II): 568–575.

[14] MALETIĆ S, STAMENIĆ L, RAJKOVIĆ M. Statistical mechanics of simplicial complexes [J]. Atti Semin. Mat. Fis. Univ. Modena Reggio Emilia, 2011, 58: 245–261.

[15] MALETIĆ S, RAJKOVIĆ M. Combinatorial Laplacian and entropy of simplicial complexes associated with complex networks [J]. Eur. Phys. J. Special Topics, 2012, 212: 77.

[16] MALETIĆ S, HORAK D, RAJKOVIĆ M. Cooperation, conflict and higher-order structures of complex networks [J]. Advances in Complex Systems, 2012, 15: 1250055.

[17] NEWMAN M E J. Networks: An introduction [M]. Oxford: Oxford University Press, 2010.

[18] CALDARELLI G. Scale-free networks: Complex webs in nature and technology [M]. Oxford: Oxford University Press, 2007.

[19] LOVÁSZ L. Kneser's conjecture, chromatic numbers and homotopy [J]. J. Comb. Th. A, 1978, 25: 319.

[20] ARENAS F G, PUERTAS M L. The Neighborhood complex of an infinite graph [J]. Divulgaciones Matematicas, 2000, 8: 69.

[21] JOHNSON J H. Some structures and notation of Q-analysis [J]. Environment and Planning B, 1981, 8: 73.

[22] DEGTIAREV K Y. Systems analysis: Mathematical modeling and approach to strucutral complexity measure using polyhedral dynamics approach [J]. Complexity International, 2000, 7: 1.

[23] BRON C, KERBOSCH J. Finding all cliques of an undirected graph [J]. Comm. ACM, 1973, 16: 575.

[24] KOZLOV D. Combinatorial algebraic topology [M]. Heidelberg: Algorithms and Computation in Mathematics, Springer-Verlag, 2008.

[25] GOLDBERG T E. Combinatorial Laplacians of simplicial complexes [M]. New York: Annandale-on-Hudson, 2002.

[26] BANERJEE A, JOST J. Graph spectra as a systematic tool in computational biology [J]. Discrete Applied Mathematics, 2009, 157: 2425.

[27] MALETIĆ S, ZHAO Y, RAJKOVIĆ M. Persistent topological features of dynamical systems [J]. Chaos., 2016, 26: 053105.

[28] IKEDA K. Multiple-valued stationary state and its instability of the transmitted light by a ring cavity system [J]. Opt. Commun., 1979, 30: 257.

[29] IKEDA K, DAIDO H, AKIMOTO O. Optical turbulence: Chaotic behavior of transmitted light from a ring cavity [J]. Phys. Rev. Lett., 1980, 45: 709.

[30] RÖSSLER O. An equation for continuous chaos [J]. Physics Letters, 1976, 57A (5): 397.

[31] MULDOON M R, MACKAY R S, HUKE J P, *et al.* Topology from time series [J]. Physica D, 1993, 65: 1.

[32] GARLAND J, BRADLEY E, MEISS J D. Exploring the topology of dynamical reconstructions [J]. arXiv:1506.01128v1.

[33] EDELSBRUNNER H, HARER J L. Computational topology: An introduction [M]. Providence: American Mathematical Society, 2010.

[34] GRASSBERGER P, PROCACCIA I. Measuring the strangeness of strange attractors [J]. Physica D, 1983, 9: 189.

[35] FRASER A M, SWINNEY H L. Independent coordinates for strange attractors from mutual information [J]. Phys. Rev. A, 1986, 33: 1134.

[36] CARLSSON G. Topology and data [J]. American Mathematical Society Bulletin, 2009, 46(2): 255.

[37] MOGFORD R H. Mental models and situation awareness in air traffic control [J]. International Journal of Aviation Psychology, 1997, 7(4).

[38] ZANIN M, LILLO F. Modelling the air transport with complex networks: A short review [J]. Eur. Phys. J. Special Topics, 2013, 215: 5.

[39] LORDAN O, SALLAN J M, SIMO P. Study of the topology and robustness of airline route networks from the complex network approach: A survey and research agenda [J]. Journal of Transport Geography, 2014, 37: 112.

[40] PASTO-SATORRAS R, VESPIGNANI A. Epidemic spreading in scale-free networks [J]. Phys. Rev. Lett., 2001, 86(14): 3200.

[41] COLIZZA V, BARRAT A, BARTHÉLEMY M, *et al.* The role of the airline transportation network in the prediction and predictability of global epidemics [J]. PNAS, 2006, 103(7): 2015.

[42] MURUGESAN M, SUJITH R I. Combustion noise is scale-free: Transition from scale free to order at the onset of thermoacoustic instability [J]. Fluid Mech, 2015, 772: 225–245.

[43] PAVLOV D, MULLEN P, TONG Y, *et al.* Structure-preserving discretization of incompressible fluids [J]. Physica D: Nonlinear Phenomena, 2011, 240(6): 443.

5

Take-home Messages

In this last chapter we will give a short overview of the contents of the book and emphasis the main conclusions, as well as to propose a brief list of books and research papers for readers who want to learn more about simplicial complexes and algebraic topology.

5.1 Brief book summary

We began the story about simplicial complexes by defining them in three equivalent ways, hence inkling from the beginning on the possible versatile applications. Interestingly, defined in different ways, simplicial complexes represent mathematical objects which can be characterized with the same set of quantities. Put it in another words, although simplicial complex can be defined as geometrically or as abstract or from the relation between two sets, we can calculate the same quantities on each of them.

Within the framework of Q-analysis we have introduced structural vectors for characterization of mesoscopic structures, that is the aggregations of simplices at different dimensional levels, whereas for the characterization of the local properties of simplices we have introduced eccentricity and vertex significance. For global, macroscopic, characterization of simplicial complex we have introduced the simplicial structural complexity. Covering characterization at all three structural scales (local, mesoscopic, and global) the Q-analysis proved as powerful tool for the analysis of complex system represented as simplicial complex.

Whereas in the Q-analysis description of simplicial complexes the high-dimensionality of simplices was of great importance, in the homological calculations the attention is focused on the non-bounding cycles of simplices which does not have to be necessarily high-dimensional. Hence, we have shown that the non-bounding cycles of simplices actually make holes in the structure of simplicial complexes, and further that they form the homology group. The number of generators of q-th homology group is called q-th Betti number and represents an important topological invariant. After introducing homology groups and topological invariants related to it, we have extended the notion of homology in order to capture the homological changes of a space that is undergoing growth, and hence introduced the persistent homology. In the essence of the persistent homology lies the idea of detecting changes in topological space which are encoded in the homology groups at different stages of change. In other words, we want to follow the changes of homology groups during the changes of simplicial complex, and the changes of simplicial complex are usually followed by tuning some free parameter. With homological properties of simplicial complex we have related the q combinatorial Laplacian, which can be understood as a generalization of the graph Laplacian to higher-dimensional structures, or reversely, the graph Laplacian can be understood as the 0-th combinatorial Laplacian. Nevertheless, the eigenvalue spectra of combinatorial Laplacian carries more information about mesoscopic structures of simplicial complex than just relating the number of eigenvalues equal to zero to the Betti numbers. The versatile applications and properties of the combinatorial Laplacian are yet to be revealed.

When we set the stage of mathematical tools for characterization of simplicial complexes, we proceeded to the introduction of ways of building the simplicial complexes from data. It turned out that, although, the building blocks of a graph are rather simple (i.e., nodes and links) we can build different, and rich in structure, simplicial complexes. Hence, we defined the clique complex, the neighborhood complex, the independence complex, and the matching complex, and each of them can be built from a single graph, providing us with useful information about the graph's higher-dimensional structures.

Next we introduced different ways to build simplicial complex from the data embedded in metric space. Since the distances between elements (i.e., points or vertices or nodes) are of essential importance they play the crucial role in building simplicial complexes: the Čech complex, the Vietoris-Rips complex, and the witness complex. Finally, we introduced the building of the simplicial complex from time series. As it turned out, for that procedure we had to combine previously introduced tool (the persistent homology) and the way of building the simplicial complex from data in metric space (the Čech complex). Hence, we demonstrated how combining different algebraic topology concepts the new methods and concepts can emerge.

In order to illustrate the diversity of practical and theoretical applications of simplicial complexes we considered three cases. The opinion exchange model is introduced as an example of the application in computational social science, with the emphasis that representing an opinion as a simplex and collection of different opinions as simplicial complex emerges as natural setting for modeling social climate. And furthermore, due to the definition of an opinion space as a simplicial complex built from the relation between two sets, the opinion exchange model is easily adaptable to the modeling of public opinion surveys through the adaptation of questionnaires to the two-set definition of simplicial complex. Building different simplicial complexes (the neighborhood complex, the clique complex, and the conjugate clique complex) from single complex network (the Air Traffic Control network) showed the abundance of various substructures which are embedded in the network. And this example also illustrates the diversity of results when both Q-analysis and homology are applied, which supports the course of research which transcends the application of graph-theoretic concepts in the understanding the mesoscopic structures of complex networks. In the last example the emphasis was given to the specific application of the persistent homology in the reconstruction of phase space of dynamical system when the data of only one variable is known. As it turned out, the topological properties of reconstructed simplicial complex retain the topological properties of the original manifold of dynamical system.

The generality of all three examples provides the applicability in diverse research fields. Specifically, we emphasized the possible applications on the research in aeronautical science, nevertheless the practitioners and researchers involved in the field of aeronautics can find many different applications.

Finally, we would like to point that all examples are devised in a way such the reader can work them again in order to easy understand the abstract concepts and to understand the computational technique. Actually, our intention was to start with simple examples and gradually make them more complicated so that they support the gradual acquiring the knowledge.

5.2 Suggestions for further reading

A multitude of references which appear in this book may leave the interested reader a bit confused when he or she wants to expand and deepen the knowledge about the topics presented. In this final section we will give a short list of books and research articles which can be useful from our subjective point of view. Of course, we do not insist that this is the best choice, but rather part of the much broader list.

Algebraic topology:

- J. R. Munkres: *Elements of Algebraic Topology*, Addison-Wesley Publishing, California, 1984.
- A. Hatcher: *Algebraic Topology*, Cambridge University Press, Cambridge, 2002.
- D. Kozlov: *Combinatorial Algebraic Topology*, Algorithms and Computation in Mathematics, Springer-Verlag, Berlin Heidelberg, 2008.
- H. Edelsbrunner and J. L. Harer: *Computational topology: An Introduction*, American Mathematical Society, Providence, RI, 2010.

Q-analysis:

- R. H. Atkin: *From cohomology in physics to q-connectivity in social sciences*, Int. J. Man-Machine Studies 4, 341, 1972.

- R. H. Atkin: *Combinatorial Connectivities in Social Systems*, Birkhäuser Verlag, Base und Stuttgart, 1977.
- J. H. Johnson: *Some structures and notation of Q-analysis*, Environment and Planning, B 8, 73, 1981.
- P. Gould, J. Johnson, G. Chapman: *The Structure of Television*, Pion Limited, London, 1984.
- J. H. Johnson: *Hypernetworks in the Science of Complex Systems*, Imperial College Press, London, 2013.

Persistent homology:

- R. Ghrist: Barcodes: *The persistent topology of data*, Bull. Amer. Math. Soc. (N.S.) 45(1), 61, 2008.
- H. Edelsbrunner and J. Harer: *Persistent homology — a survey*, In Surveys on discrete and computational geometry, volume 453 of Contemp. Math., pages 257–282. Amer. Math. Soc., Providence, RI, 2008.
- G. Carlsson: *Topology and Data*, American Mathematical Society, Bulletin, New Series Volume 46, Number 2, 255-08, 2009.

Combinatorial Laplacian:

- T. E. Goldberg: *Combinatorial Laplacians of Simplicial Complexes*, Annandale-on-Hudson, New York, 2002.

Complex networks:

- R. Albert, A.-L. Barabási: *Statistical mechanics of complex networks*, Rev. Mod. Phys. 74, 47, 2002.
- M. E. J. Newman: *Networks: An Introduction*, Oxford University Press, USA, 2010.
- Reuven Cohen, Shlomo Havlin: *Complex Networks: Structure, Robustness and Function*, Cambridge University Press, Cambridge, 2010.

**The applications of algebraic topology in physics
can be found in:**

- T. Frankel: *The Geometry of Physics: An Introduction*, Cambridge University Press, Cambridge, 1997.
- H. Eschrig: *Topology and Geometry for Physics*, Lecture Notes in Physics, 822 Springer-Verlag Berlin Heidelberg 2011.

Index

Printed in the United States
by Baker & Taylor Publisher Services